Robert Geroch

Infinite-Dimensional Manifolds

1975 Lecture Notes

With 65 Figures

MINKOWSKI
Institute Press

Robert Geroch
Enrico Fermi Institute
University of Chicago

Cover: Lecture notes are often written in similar environments

ISBN: 978-1-927763-15-5 (softcover)
ISBN: 978-1-927763-16-2 (ebook)

Minkowski Institute Press
Montreal, Quebec, Canada
http://minkowskiinstitute.org/mip/

For information on all Minkowski Institute Press publications visit our website at http://minkowskiinstitute.org/mip/books/

Contents

1. Introduction

A manifold is, in general terms, "a space which, locally, looks like some simple space, although it may, in the large, look quite different from that simple space". Different choices of this "simple space" lead to different types of manifolds. By far the most common such choice is $\mathbb{R}^n$ (the finite-dimensional vector space consisting of n–tuples of real numbers). In this case, the corresponding manifolds are said to be finite-dimensional (specifically, n–dimensional), and the study of such manifolds and their structure is called differential geometry.

Finite-dimensional manifolds are important in both mathematics and physics. In mathematics, the study of finite-dimensional manifolds, differential geometry, is an end in itself. However, there exist in addition applications of manifolds to other areas of mathematics. An outstanding example is that of ordinary differential equations. The question of the existence, uniqueness, and structure of the solutions of ordinary differential equations can be reduced to the study of vector fields and their integral curves on manifolds. In this way, one recasts the subject of ordinary differential equations into an elegant, geometrical, and remarkably simple form. In physics, for example, the space-time of general relativity is found to have the structure of a finite-dimensional (in fact, four-dimensional) manifold. As a second example from physics, the space of configurations of a mechanical system becomes a manifold (where the dimension is what is called the number of degrees of freedom of the system).

One can also introduce manifolds based on spaces which are "larger than finite-dimensional vector spaces", i.e., on "infinite-dimensional spaces". Although it is perhaps true that the possible applications of such manifolds have not yet been exploited fully, there are already clear indications that these applications will be rich and far-ranging. One may expect, for example, that the subject of partial differential equations (particularly, hyperbolic and parabolic equations) can be formulated in the same geometrical language as was possible for ordinary differential equations. In physics, infinitive-dimensional manifolds have already found their way into such areas as general relativity and quantum field theory.

One might, therefore, wish to learn about infinite-dimensional manifolds. The next issue one must resolve is this: On what "simple (but infinite-dimensional) spaces" should these manifolds be based? There is only one such choice for which, as far as I am aware, the subject has been worked out in any detail, namely, that in which the simple space is what is called a Banach space. It is my own view that manifolds based on Banach spaces might be "too right in structure" for certain applications. That is to say, although one obtains a large number of strong theorems in this case, it may sometimes turn out to be the case that it is not possible to make the space one is considering in some application into a manifold based on a Banach space. Nonetheless, one has to begin this subject somewhere, and it seems advisable to begin in that regime in which i) many theorems are available, and ii) the subject has been worked out in detail. If one should later decide that other types of manifolds will be more fruitful, then one will at least have a feeling for the kinds of things which are and are not likely to be true, the techniques which are likely to lead to proofs, and where to look for counterexamples.

We shall here study manifolds based on Banach spaces: their structure, and the kinds of objects which can be placed on them.

2. Banach Spaces

In this section, we define a Banach space, give a few examples and introduce related notions.

A *Banach space* consists of two things:

1. A real vector space E. That is to say, E is a set (whose elements are called *vectors*), together with a rule which assigns to any two vectors another (called their *sum*), and another rule which assigns to any vector and any real number a vector (called the *product* of the number and the vector), subject to the usual conditions for a vector space (namely, addition is commutative and associative, there is an additive identity, there are additive inverses, multiplication of vectors by numbers is distributive, and $1 \cdot x = x$ for any vector x).

2. A norm on the vector space E. That is to say, we are given a mapping from the vector space E to the reals. The real number which is the image of the vector x under this mapping is written $|x|$, and is called the *norm* of x. This mapping must satisfy the following three conditions: i) for any vector x, $|x| \geq 0$, with equality if and only if $x = 0$, ii) for any two vectors x and y, $|x + y| \leq |x| + |y|$, and iii) for any vector x and any number a, $|ax| = |a|\,|x|$. [These three conditions are perhaps rather natural if one interprets the norm of a vector as its "length".]

Finally, in order that $E, |\ |$ be a Banach space, it is necessary that it be complete, as described below.

Let E be a real vector space, with norm $|\ |$. Let $x_1, x_2, \ldots$ be a sequence of vectors in E. This sequence is called a *Cauchy sequence* if the following property is satisfied: given any positive number ϵ, there exists an integer N such that $|x_i - x_j| \leq \epsilon$ whenever $i \geq N$ and $j \geq N$. [In intuitive terms, the elements of a Cauchy sequence "get closer and closer to each other along the sequence." This sequence is said to *converge* to vector x if the limit of $|x - x_j|$ as j goes to infinity is zero (i.e., in intuitive terms, if "the elements of the sequence get closer and closer to x"). It is easily checked that a sequence that converges to some x is automatically a Cauchy sequence. Our $E, |\ |$ is said to be *complete* if the converse also holds therein, i.e., if every Cauchy sequence in $E, |\ |$ converges to some vector in E.

Thus, a Banach space is a complete normed vector space. We now give some examples.

Example. Let E be the vector space of real numbers (so addition of vectors is addition of real numbers; and multiplication of vectors by numbers is multiplication of real numbers). Let the norm of vector (real number) x be the absolute value. The three properties of the norm are elementary facts about the absolute value; completeness is a basic fact about the real number system. Thus, we obtain a Banach space.

Example. Let E be any finite-dimensional vector space. Choose any subset B of E which has the following two properties: i) B is convex, i.e., given any vectors x and y in B, and any number a with $0 \leq a \leq 1$, the vector $ax + (1 - a)y$ is in B, and ii) B is radial, i.e., given any non-zero vector x, there is a number $a_0 > 0$ such that ax is in B whenever $0 < a < a_0$ and ax is not in B whenever $a > a_0$. [The first property requires that "the line segment joining any two vectors in B lies entirely within B"; the second requires that "any ray emanating from 0 is within B for a while, and then leaves and remains outside of B".] We use this B to define a norm on E as follows; for x any nonzero vector in E, set $|x| = 1/a_0$, where a_0 is the number in property ii) above. Then the first and third properties for a norm follow from the fact that B is radial, while the second follows from the fact that B is convex. Completeness is easily checked. Thus, we obtain a Banach space. [For example, if B is the "unit sphere" about 0, then the norm of a vector is just its "Euclidean length".]

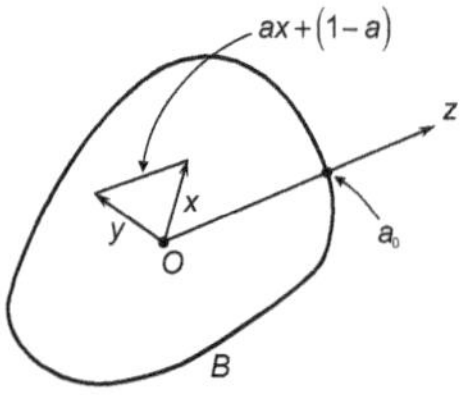

Indeed, it is not difficult to show that this last example yields the "generic" finite-dimensional Banach space, in the sense that every such Banach space arises as in the example. [Sketch of proof: Given a finite-dimensional space, $E, |\ |$, denote by B the set of all vectors x with $|x| < 1$. This set B satisfies the two conditions above, and generates, by the construction above, the original norm on E.]

Example. Denote by E the collection of all sequences of real numbers, $(r_1, r_2, \ldots)$, which are bounded. Define addition and multiplication by numbers component-wise (*i.e.*, $(r_1, r_2, \ldots) + a(r'_1, r'_2, \ldots) = (r_1 + a r'_1 \cdot r_2 + a r'_2, \ldots))$. Then E is a vector space. Next, define a norm on E as follows: $(r_1, r_2, \ldots = \text{lub}|r_1|$ (noting that the least upper bound on the right exists, since the r_i must be bounded for membership in E). That this is indeed a norm is an easy check. Finally, one must check completeness. To this end, let $x = (r_1, r_2, \ldots)$, $x' = (r'_1, r'_2, \ldots)$, $x'' = \cdots$ be a Cauchy sequence of elements of E. Then the sequence of real numbers, $r_1, r'_1, r''_1, \ldots$ is Cauchy, whence it converges to some real number s_1; the sequence $r_2, r'_2, \ldots$ of real number is Cauchy, whence it converges to some real number s_2: etc. In this way, one constructs a sequence $(s_1, s_2, \ldots)$. Since $(r_1, r_2, \ldots)$ is bounded, and since $x, x', x'', \ldots$ is Cauchy, the sequence $(s_1, s_2, \ldots)$ s bounded: hence, it defines some element y of E. Finally, one observes that, by the construction by which y was obtained, the sequence $x, x', x'', \ldots$ in E converges to y. Thus, we have sketched the proof that E is complete: hence, it is a Banach space.

Example. If, in the example above, one had restricted E to consist only of sequences $(r_1, r_2, \ldots)$ for which $\lim r_i$ (as $i \to \infty$) exists, using the same norm as above, then one would again have obtained a Banach space. Similarly, one could have restricted to $(r_1, r_2, \ldots)$ with $\lim r_1 = 0$, and would still obtain a Banach space. If, however, one had let E consist of sequences $(r_1, r_2, \cdots)$ with the property that, for some N, $r_i = 0$ for $i \geq N$, one would still obtain a vector space E, and one would still have the above norm. However, this would not be a Banach space, for it would not be complete. [For example, let $x = (1, 0, 0, \ldots)$, $x' = (1, 1/2, 0, 0, \ldots)$, $x'' = (1, 1/2, 1/4, 0 \ldots)$, etc, Then $x, x', x'', \cdots$ is a cauchy sequence, but converges to no element of this E (for $(1, 1/2, 1/4, 1/8, \ldots)$ is not in this case an element of E).]

Example. Let E consist of all bounded, real-valued functions on the reals. For f such a function, let $|f| = \text{lub}|f|$. This is a vector space (where addition of functions and multiplication by numbers are point-wise), with norm. In fact, this is a Banach space. [The proof of completeness is identical with that at the bottom of previous page.] If E had consisted only of the continues bounded functions, we would still have a Banach space. Suppose, however, that E had consisted of the differentiable bounded functions. Then, although we would still have a vector space norm, completeness would fail, and we would not obtain a Banach space.

Indeed, consider the sequence of differentiable functions illustrated on the right. These are a Cauchy sequence, but it converges to no differentiable function (the only candidate being the non-differentiable function shown).

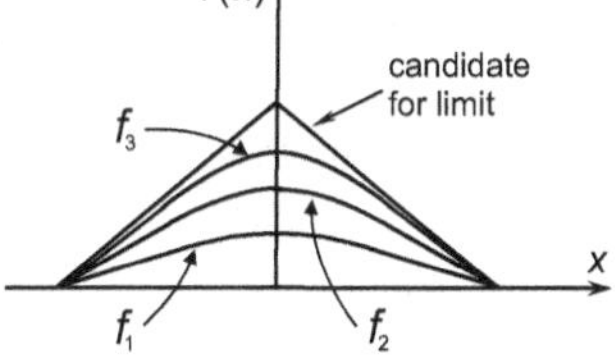

Example Let E consist of all C^n, real-valued functions on the reals, such that the values of the function and of its first n derivatives are all bounded. Let the norm of such a function be $f = \text{lub}|f| + \text{lub}|f'| + \ldots + \text{lub}|f^n|$. This is a Banach space. [If the last term were left off the expression for the norm, completeness would fail.]

Example. Let E consist of all sequences, $(r_1, r_2, \ldots)$, of real numbers, for which the sum $\Sigma|r_i|$ is finite. Let the norm of such a sequence be this sum. This is a Banach space. More generally, fixing any numbers $p \geq 1$, let E consists of $(r_1, r_2, \ldots)$ with $\Sigma|r_i|^p$ finite. Let the norm of an of an element of E be the p–th root of this sum. [The taking of root is necessary in order to have $|ax| = |a||x|$.] We obtain a Banach space.

There is apparently no natural way to make spaces of C^∞ functions into

Banach spaces, although there is for analytic functions. It is important that one understand the examples above completely.

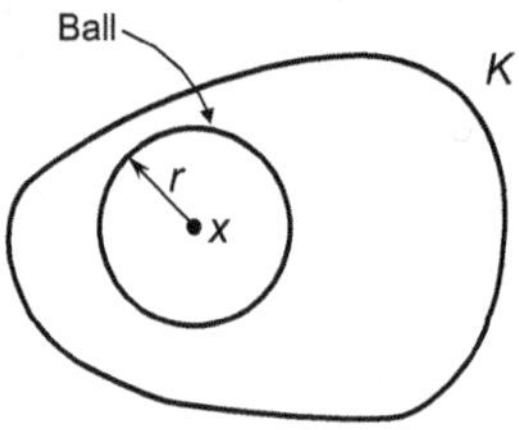

We next consider a few additional definitions associated with Banach spaces. Let E, $|\ |$ be a Banach space. For x any vector in E, and r any positive number, the *ball* with *center* x and *radius* r is the subset of E consisting of all vectors y with $|x - y| < r$. The intuitive meaning is that suggested by these terms. [For example, in the second example on 4, the ball with center 0 and radius 1 is essentially the set B of that example.] A subset K of our Banach space is said to be *open* if, given any vector x in K, there is some ball with center x which lies entirely within K. For example, every ball is itself open (a fact whose proof requires use of the second property of a norm). [Those familiar with topology will notice that these are the open sets for a topology on E.] A subset K of E is said to be *closed* if its component in E is open, i.e., if, given any vector x not in K, there is some ball centered at x which does not intersect K.

Now let E be a vector space, and let $|\ |$ and $\{\ \}$ be two norms on E, such that both E, $|\ |$ and E, $\{\ \}$ are Banach spaces. These two norms will be said to be *equivalent* if there are positive numbers a and b such that, for every vector x, $|x| \leq a\{x\}$ and $\{x\} \leq b|x|$. [In intuitive terms, if each norm, after a suitable rescaling, bounds the other.] Note that equivalence of norms is an equivalence relation, and that two norms are equivalent if and only if they define precisely the same open sets on E. It normally happens in practice that the actual numerical values associated with a norm are not relevant to one's particular problem, but rather only that norm up to equivalence. Indeed, it is possible to treat our subject referring only to the open sets (i.e., to equivalence classes of norms), rather than to any particular norm. The resulting treatment is more elegant, and occasionally more awkward. We shall not proceed in this way, however, because it would require a brief retour into topology. As an example of equivalence, we may note the following fact: any two $B's$ in the second example on page 4 give rise to equivalent norms.

Let E, $|\ |$ be a Banach space. A subset F of E is called a *subspace* if the following two conditions are satisfied: i) F is a vector subspace of vector space E (i.e., sums and numerical multiplies of vectors in F are again in F), and ii) F is a closed subset of E. As motivation for this definition, we make the following observation. Let F be a subspace of Banach space E. Then, by condition i), F is itself a vector space. Since every vector in F is also a vector in E, the norm on E also defines a norm on F. We now claim that condition ii) implies that this vector space F, with this norm, is complete (i.e., is a Banach space). [Sketch of proof: Since the norm in F comes from that in E, every Cauchy sequence in F is also a Cauchy sequence in E. Since E is complete, every such sequence converges to some vector in E. Since F is a closed subset of E, this vector must in fact be in F. Hence, F is complete.]

We see here the first of several differences we shall encounter between finite-dimensional and infinite-dimensional Banach spaces. In the finite-dimensional case, condition ii) for a subspace follows already from condition i), i.e., every vector subspace of finite-dimensional E is necessary closed. [Geometrically, e.g., a plane through the origin in Euclidean 3−space is necessarily closed.] This property does not hold, however, for infinite-dimensional Banach spaces:

Example. Let E be the Banach space of the example on page 4 (bounded sequences of real). Let F be the subset of E consisting of all sequences, $(r_1, r_2, \ldots)$ such that $r = 0$ for all $i \geq N$ for some N. [That is, F consists of sequences which are "all zeroes after a while".] Then, since the sum of two sequences in F is again in F, and since a numerical multiple of a sequence in F is again in F, F is a vector subspace of vector space E. We claim, however, that this subset F is not closed. Consider the element $x = (1, 1/2, 1/4, \ldots)$ of E. This x is not an element of F. If F is to be closed, then there must exist a ball, centered at x, which fails to intersect F. We show that there is no such ball, whence F cannot be closed. Let r be any positive number, and let B be the ball centered at x, with radius r. We must find a vector common to B and F. Set $y = (1, 1/2, 1/4, \ldots, 1/2^n, 0, 0, \ldots)$, where n is chosen sufficiently large that $1/2^n \leq r$. Then $|x - y| = (0, 0, \ldots, 0, 1/2^{n+1}, 1/2^{n+2}, \ldots)$, whence $|x - y| = 1/2^{n+1}$. Thus, the vector y is in the ball B. But y is also in F (since F consists of sequences which are 0 after a while). Hence, this y is common to B and F. [Compare this argument with the argument, in the first part of page 5 that F itself is not a Banach space.]

3. Banach Spaces from Banach Spaces

In this section, we discuss two methods for constructing Banach spaces from Banach spaces. In particular, applied to our examples these methods yield numerous new examples.

Let E and F be Banach spaces. Denote by G the collection of all ordered pairs, (x, y), with x a vector in E and y in F. The set G is made into a vector space by defining addition and scalar multiplication component-wise, i.e., $(x, y) + a(x', y') = (x + ax', y + ay')$. We now wish to make this vector space G into a Banach space, i.e., we wish to define a norm on G. Set $|(x, y)| = |x| + |y|$. The three conditions for a norm are immediate from those conditions in E and F: e.g., $|a(x, y)| = |(ax, ay)| = |ax| + |ay| = |a|\,|x| + |a|\,|y| = |a|(|x| + |y|) = |a||(x, y)|$. Thus, to show that G is a Banach space, we must only check completeness. Let $(x_1, y_1), (x_2, y_2), \ldots$ be a Cauchy sequence in G. Then, for every positive ϵ there is an N such that $|(x_i, y_i) - (x_j, y_j)| \leq \epsilon$ whenever $i, j \geq N$. But $|(x_i, y_i) - (x_j, y_j)| = |x_i - x_j| + |y_i - y_j|$. Hence, $x_1, x_2, \ldots$ is a Cauchy sequence in E (whence it converges to some vector x in E), and $y_1, y_2, \ldots$ is a Cauchy sequence in F (whence it converges to some vector y in F). Consider now the vector (x, y) in G. We have $|(x, y) - (x_i, y_i)| = |x - x_i| + |y - y_i|$. Since $\operatorname{limit}_{i\to\infty} |x - x_i| = \operatorname{limit}_{i\to\infty} |y - y_i| = 0$ (convergence in E and F), we have $\operatorname{limit}_{i\to\infty} |(x, y) - (x_i, y_i)| = 0$. Hence, our sequence converges to (x, y). We have shown that G is complete; hence, is a Banach space. This G is called the *product* of Banach spaces E and F, written $G = E \times F$, Note, for example, that each of E and F may be regarded as a subspace of $E \times F$.

The second construction is like an "inverse" of the first. Let E be a Banach space, and let F be a subspace of E. Then in particular F is a vector subspace of vector space E, and so we may take the quotient of vector spaces; denote the resulting vector space by G. [In more detail: take two vectors x and x' of E as equivalent if their difference, $x - x'$, is an element of subspace F. This is an equivalence relation, and G is the set of equivalence classes. To add two elements of G (equivalence classes), find the equivalence class which includes the sum of one representative from each summand (noting that this is independent of the choice of representative), and similarly for scalar multiplication.] We next introduce a norm on this vector space G. For α any element of G (equivalence class), let $|\alpha|$ be the greatest lower bound of $|x|$, as x varies over the equivalence class α. This is indeed a norm on G. [The third condition for a norm is immediate, and the second is not hard. For the first, first note that always $|\alpha| \geq 0$. If $|\alpha| = 0$, then there is a sequence $x_1, x_2, \ldots$ in α with limit $|x_i| = 0$. Since F is closed, therefore, the vector 0 in E must be in α. Hence, since α has representative 0, we have $\alpha = 0$.] Finally, we show that this G is complete. Let $\alpha_1, \alpha_2, \ldots$ be a Cauchy sequence in G: By choosing a subsequence if necessary, we may suppose that $|\alpha_{i+1} - \alpha_i| \leq 1/2^i$. Choose any vector x_1 in α_1. Choose vector x_2 in $\alpha_2 - \alpha_1$ with $|x_2| \leq |\alpha_2 - \alpha_1| + 1/2$, vector x_3 in $\alpha_3 - \alpha_2$ with $|x_3| \leq |\alpha_3 - \alpha_2| + 1/4$, etc. [These choices are possible by definition of the norm in G.] Thus, for $i \geq 2$ we have $|x_i| \leq 1/2^{i+2}$. Next, set, for each $i, y_i = x_i + x_2 + \ldots + x_i$. Then, for each i, y_i is in α_i. Further, by the bounds above on the norms of the x's, the y_i form a Cauchy sequence in E. Hence, this sequence converges to some vector y in E. Denote by α the equivalence class containing y. Then $|\alpha - \alpha_i| \leq |y - y_i|$ (by definition of the norm in G), and so, since the y_i converge to y in E, the α_i must converge to α in G. We have shown that every Cauchy sequence in G converges to some vector in G, i.e., that G is complete. This Banach space G is called the *quotient* of Banach space E by subspace F, and is written E/F.

The sense in which these two constructions invert each other is this. Let E and F be Banach spaces. Then E is a subspace of the Banach space $E \times F$ (namely, the subspace consisting of pairs (x, y) with $y = 0$). Hence, we may take the quotient of $E \times F$ by this subspace E. The result is what one expects; $(E \times F)/E = F$

4. Open Mapping Theorem

In this section we establish an important property of Banach spaces. This property is not only useful in applications, but also gives insight into the structure of the definition of a Banach space.

We shall regard the open mapping theorem as a criterion for equivalence of norms.

Theorem (open mapping). Let $E, |\ |$ and $E, \{\ \}$ be two Banach spaces (based on the same vector space E), and suppose that there is a positive number a such that $\{x\} \leq a|x|$ for every vector x. Then these two norms are equivalent, i.e., also $|x| \leq b\{x\}$ for some b.

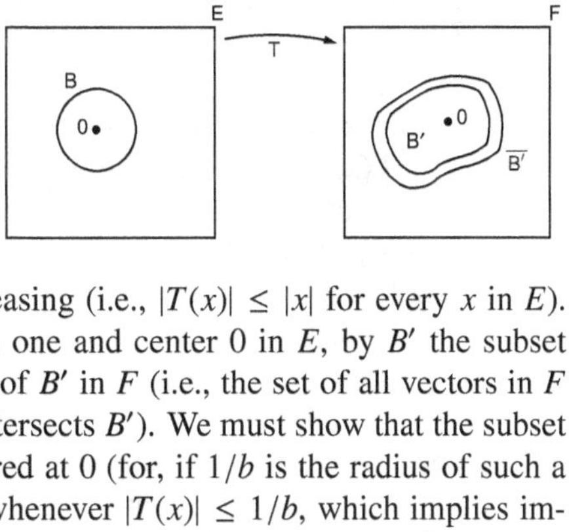

Proof: First note that no generality is lost by setting $a = 1$. Next, introduce a copy, $F, |\ |$, of Banach space E, $\{\ \}$, and let T denote that mapping from E to F which arises from the identity mapping on E. Then T is linear, one-to-one, onto, and norm-decreasing (i.e., $|T(x)| \leq |x|$ for every x in E). Denote by B the ball with radius one and center 0 in E, by B' the subset $T(B)$ of F, and by $\overline{B'}$ the closure of B' in F (i.e., the set of all vectors in F every sphere centered at which intersects B'). We must show that the subset B' of F contains some ball centered at 0 (for, if $1/b$ is the radius of such a ball, then we shall have $|x| \leq 1$ whenever $|T(x)| \leq 1/b$, which implies immediately that $|T(x)| \leq b|x|$ for every x in E). We divide the proof into three steps.

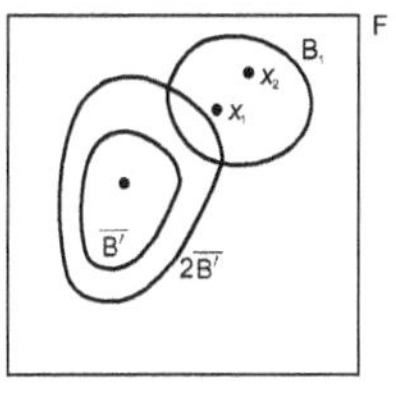

The set $\overline{B'}$ in F contain some ball. Suppose not. We obtain a contradiction. Choose vector x_1 not in $\overline{B'}$. Then, since $\overline{B'}$ is closed, there is some ball B_1, centered at x_1, whose closure $\overline{B_1}$ does not intersect $\overline{B'}$. The set $2\overline{B'}$ (the set of all vectors of the form $2x$ with x in $\overline{B'}$) cannot contain a ball (since $\overline{B'}$ does not), and hence in particular cannot contain B_1. Choose, therefore, vector x_2 in B_1 but not in $2\overline{B'}$. Since $2\overline{B'}$ is closed, there is some ball B_2 centered at x_2 such that $\overline{B_2}$ does not intersect $2\overline{B'}$. Choose B_2 a subset of B_1, and with radius less than half that of B_1. Now $3\overline{B'}$ cannot contain B_2: choose x_3 in B_2 but not in $3\overline{B'}$. Then, choose ball B_3 centered at x_3, not intersecting $3\overline{B'}$, a subset of B_2, and having radius less than half of that of B_2. Continuing in this way, we obtain a sequence $x_1, x_2, \ldots$ in F, which, by construction, must be a Cauchy sequence. Hence, this sequence converges to some vector x in F. By construction, this x cannot be in $\overline{B'}$ (since $\overline{B_1} \cap \overline{B'} = \emptyset$) or in $2B'$ (since $\overline{B_2} \cap 2\overline{B'} = \emptyset$), etc. But this is a contradiction, for $\overline{B'} \cup 2\overline{B'} \cup 3\overline{B'} \ldots = F$ (which follows from the fact that $B \cup 2B \cup 3B \cup \ldots = E$, and that T is onto).

The set $\overline{B'}$ in F contains some ball centered at 0. Let $\overline{B'}$ contain a ball centered at $T(x)$, where x is some vector in E. Then $\overline{B'} - T(x)$ contains a ball centered at 0. But $\overline{B'} - T(x) = \overline{T(B)} - T(x) = \overline{(T(B) - T(x))} = \overline{T(B - x)}$. Choose positive number n such that $B - x$ is a subset of nB. Then $\overline{T(nB)} = n\overline{T(B)} = n\overline{B'}$ contains a ball centered at 0 (since $\overline{T(B - x)}$ does). Hence, $\overline{B'}$ also contains a ball centered at 0.

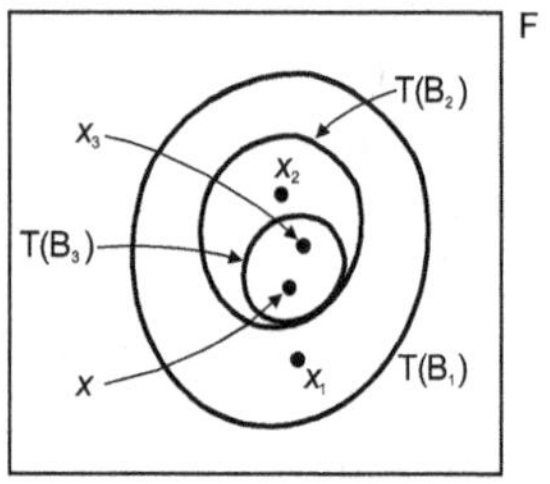

The set B' in F contains some ball centered at 0. Suppose not: We obtain a contradiction. Let $\overline{B'}$ contain the ball of radius ϵ centered at 0. Since B' contains no ball centered at 0, neither does $3B'$. But $\overline{B'}$ does contain a ball centered at 0: Hence, we may choose a vector $T(x)$ in $\overline{B'}$, but not in $3B'$. Since $T(x)$ is in $\overline{B'}$, we may choose a vector $T(x_1)$ in B' with $|T(x) - T(x_1)| \leq \epsilon/4$. Denote by B_1 the ball of radius $1/2$ about x_1. Then (since $|T(x) - T(x_1)| \leq \epsilon/4$ and $\overline{T(B_1)}$ contains the ball of radius $\epsilon/2$ about $T(x_1)$) $\overline{T(B_1)}$ contains the ball of radius $\epsilon/4$ about $T(x)$. Hence, we may choose vector $T(x_2)$ in $T(B_1)$ with $|T(x) - T(x_2)| \leq \epsilon/8$. Denote by B_2 the ball of radius $1/4$ about x_2. Then $\overline{T(B_2)}$ contains a ball of radius $\epsilon/8$ about x. Hence, we may choose vector $T(x_3)$ in $T(B_2)$ with $|T(x) - T(x_3)| \leq \epsilon/16$. Denote by B_3 the ball of radius $1/8$ about x_3. Continuing in this way, we obtain a sequence of vectors $x_1, x_2, \ldots$ in E. Since $|T(x) - T(x_i)| \leq \epsilon/2^{i+1}$, the $T(x_i)$ converge to $T(x)$. Since $T(x_i)$ is in $T(B_{i-1})$, x_i must be in B_{i-1}, i.e., we have $|x_i - x_{i-1}| \leq 1/2^{i-1}$. Hence, $x_1, x_2, \ldots$ is a Cauchy sequence in E. Denote by y the vector in E to which it converges. Then, since x_1 is in B (i.e., since $|x_1| < 1$), and since $|x_i - x_{i-1}| \leq 1/2^{i-1}$, we must have $|y| < 3$, i.e., y must be in $3B$. Furthermore, since T is norm-decreasing, and since limit $|y - x_i| = 0$, we must have limit $|T(y) - T(x_i)| = 0$, i.e., the sequence $T(x_i)$ in F converges to $T(y)$. But we have already seen that the $T(x_i)$ converges to $T(x)$: Hence, $T(x) = T(y)$. Thus, since $T(y)$ is in $3B'$, $T(x)$ must also be in $3B'$. This is a contradiction with our original choice of $T(x)$.

This completes the proof of the theorem.

The result above is beautiful and intricate. To see what it means, consider first the finite-dimensional case. Imagine that we have one norm on our vector space, and wish to construct a smaller one. One could imagine obtaining a new norm by "scaling down the original norm by a positive factor, one for each direction (i.e., for each dimension) in the vector space". If there are only a finite number of dimensions, then these "scaling factors" will have a minimum, and so the new norm will also bound (up to a factor) the old one. In the infinite-dimensional case, however, things are different. Now, one could imagine "choosing different factors for the various directions, such that these factors approach zero", thus obtaining a new norm which does not bound the old norm. Suppose, however, that one wants ones new norm to yield a Banach space: Then one must be careful about completeness, for scaling down the old norm "too much" may lead to new Cauchy sequences with nothing to converge to. Is it necessary to actually have the "scaling factors" bounded away from zero in order to avoid destroying completeness? The theorem says yes.

The open mapping theorem is somewhat analogous to a theorem in topology: Given a compact, Hausdorff topological space, then no finer and no coarser topology can be both compact and Hausdorff. The open mapping theorem says, similarly: Given a Banach space, then no larger and no smaller norm (except for equivalence) can also give a Banach space. In fact, whenever, in an argument about Banach spaces, one is tempted to try to use compactness (which is almost never available in this subject), one should try instead to apply the open mapping theorem.

5. Splitting

An important issue about a subspace of a Banach space is that of whether or not it has the property called splitting. In this section, we introduce this property, show that it is always satisfied in certain cases, and give an example in which it is not satisfied.

Let E be a Banach space, and F a subspace of E. A subspace G of E is said to be *complementary* to F if the following property is satisfied: every vector in E can be written in one and only one way as the sum of one vector in F and one in G.

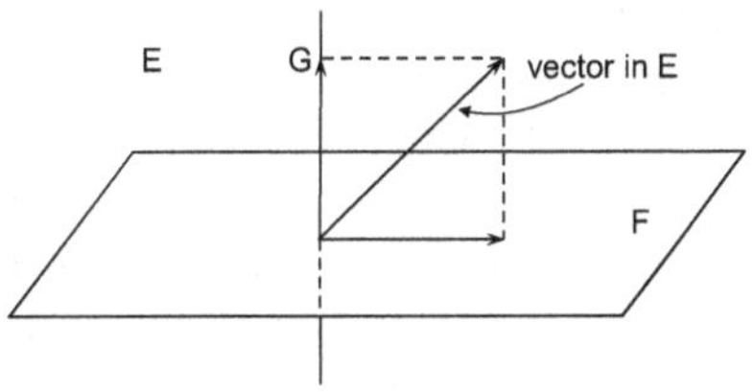

Example. Let E be the Banach space of all sequences, $(r_1, r_2, \ldots)$, of real numbers, such that the r_i converge, and let the norm of such a sequence be the least upper bound of the absolute values of its entries. Let E be the subspace consisting of sequences which converge to zero, and let G be the subspace consisting of constant sequences. Then F and G are complementary.

Let F and G be complementary subspaces of Banach space E. We define a mapping from G to the quotient (Banach) space E/F as follows: This mapping takes the vector x in G to the equivalence class (element of E/F) which contains x. This mapping is clearly linear, one-to-one, and onto. Furthermore, this mapping is norm-decreasing (since the norm of x in G is just the norm of x in E, while the norm of an equivalence class (in E/F) is the greatest lower bound of the norms of the representatives (including x) of that class). By the open mapping theorem, therefore, this mapping is an isomorphism of Banach spaces. Thus, if G is complementary to F in E, then G represents "a realization of E/F in E", The Banach space E/F starts out as just an abstract Banach space: it does not "live" in E. The finding of a complementary subspace (to F) "realizes E/F".

Let E be a Banach space, and F a subspace of E. The subspace F is said to *split* if there exists in E a subspace complementary to F. The purpose of this section is to understand this definition.

We first show that "very small" subspaces do split.

Theorem. Let F be a one-dimensional subspace of Banach space F. Then

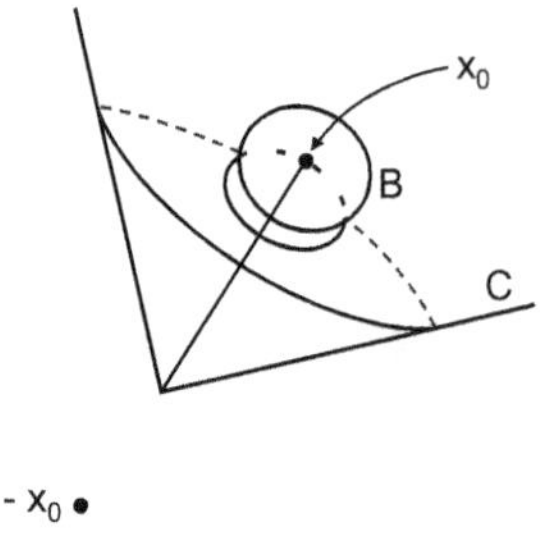

Proof: Choose vector x_0, with $|x_0| = 1$, in F. Denote by B the ball with center x_0 and radius $1/2$. Next, denote by ζ the collection of all subsets C of E having the following properties: i) C is convex (i.e., the line segment joining any two vectors in C lies within C), ii) C is radial (i.e., if x is in C and a is a positive number, then ax is in C, iii) C contains the ball B, and iv) C does not contain the vector $-x_0$. There certainly exists at least one subset in ζ, namely, that consisting of all positive multiples of vectors in B. Partially order the set ζ by inclusion (i.e., $C_1 \leq C_2$ if $C_1 \subset C_2$). We next note that this partially ordered set ζ satisfies the condition for Zorn's Lemma (namely, that every totally ordered subset of ζ is bounded above), for, given a totally ordered subset of ζ, their union gives such a bound. [Conditions i)–iv) for this union are all immediate from those conditions for the C's in the totally ordered subset.] So, by Zorn's Lemma, there exists a maximal element, C, of ζ.

We next obtain two properties of this set C. First, $C \cup (-C) = E$. Indeed, if there were some vector x in E, with x in neither C nor $-C$, then we could consider the set C', consisting of the union of all line segments joining a positive multiple of x to a vector in C. This C' would clearly satisfy conditions i), ii), and iii) above. It would, furthermore, satisfy condition iv), for if $-x_0$ were in C', say $-x_0 = ax + y$ (a positive, y in C), then we would have $x = -1/a(x_0 + y)$, whence x would be in $-C$. But the existence of this C' (an element of ζ, since it satisfies all four conditions for membership) would violate maximality of C. Hence $C \cup (-C) = E$. The second property is that C is closed. Indeed, if not, we could let C' be the closure of C. Then this C' clearly satisfies conditions i), ii), and iii) above. Furthermore, it would satisfy condition iv), for, were $-x_0$ in C', there would have to be some vector y in C with $|y + x_0| \leq 1/4$. But $-y - 1/8\,x_0$ must be in C (for $|(-y - 1/8\,x - 0) - x_0| = |y + x_0 + 1/8\,x_0| \leq |y + x_0| + |1/8\,x_0| \leq 3/8$, whence $-y - 1/8\,x_0$ is already in B). Hence, since C is convex, $1/2(y) + 1/2(-y - 1/8\,x_0) = -1/16\,x_0$ would be in C, whence $-x_0$ would be in C; a contradiction. Thus, the closure C' of C must also be an element of ζ. This violates maximality of C unless C is already closed (so $C' = C$).

Now set $G = C \cap (-C)$. This subset G of E is a vector subspace of vector space E (since, for x and y in G, $ax + (z - a)y$ is in G (for $0 \leq a \leq 1$; convexity of C and $-C$), ax is in G (for $a > 0$; radialness of C and $-C$), and $-x$ is in G (since, for x in G, x is in both C and $-C$, whence $-x$ is also in both). Furthermore, this subset G is closed (as the intersection of closed subsets C and $-C$). Thus, G is actually a subspace of Banach space E. This subspace

G is our candidate for a subspace complementary to F.

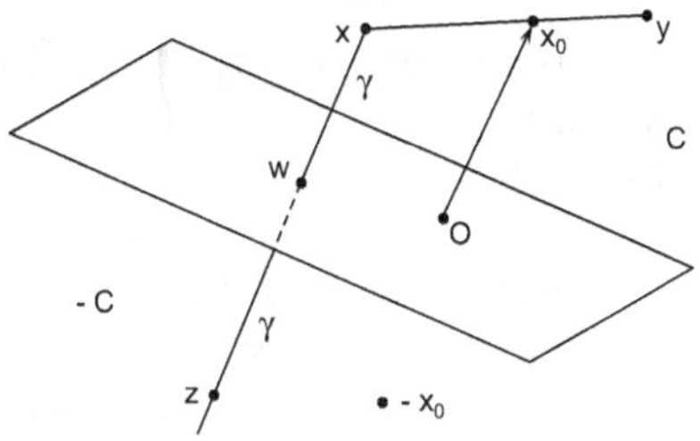

First note that no vector can be written in more than one way as a linear combination of x_0 and a vector in G, for, were this possible, x_0 itself would have to be in G (a contradiction, since x_0 is not in $-C$). Thus, we have only to show that every vector in E can be written in some way as a linear combination of x_0 and a vector in G. Let x be any vector in E (say, x in C). Denote by γ the straight line in E through x parallel to the vector x_0 (i.e., the set of all vectors of the form x plus a multiple of x_0). Choose vector y a little beyond x_0 on the line joining x and x_0. By making the "little" small enough, we can have y in B (i.e., we can have $|y - x_0| \leq 1/2$). Clearly, there is a point z on the line γ such that the line segment joining z and y contains $-x_0$. Then this z cannot be in C (for y, being in B, is necessary in C, and if z were also in C then, by convexity, $-x_0$ would be in C, contradicting definition of C). Hence, the line γ has some points in C and some in $-C$. Therefore, there must exist some point w on γ "at the interface" (i.e., such that, given any ϵ, the intersection of γ with the ball of radius ϵ about w intersects both C and $-C$). Since C and $-C$ are closed, this w must be in both C and $-C$, i.e., w must be in G. But x can be written as a linear combination of x_0 and w. Thus, the subspaces G and F are indeed complementary, completing the proof.

It is immediate from this result that every finite-dimensional subspace of a Banach space split. Indeed, let F be such a subspace, and let $x_1, \ldots, x_n$ be a basis for F. Denote by $G_1, \ldots, G_n$ a subspace complementary to that spanned by $x_1, \ldots, x_n$, respectively. Then $G \cap \ldots \cap G_n$ is a subspace complementary to F. It is also easy to show that any subspace of finite co-dimension (codimension of F in E equals dimension of R/F, i.e., is the dimension of "what is left over in E after F") splits. Indeed, choose basis $\alpha_1, \ldots \alpha_n$ for E/F, and representatives $x_1, \ldots, x_n$ or these equivalence classes in E. Let G be the subspace spanned by $x_1, \ldots, x_n$. Then, G is complementary to F. [This same argument does not work when E/F is infinite-dimensional, for then the G, so constructed, may not be closed.]

We thus conclude that, given Banach space E, "very large" subspaces (finite co-dimension) and "very small" subspaces (finite dimension) split. Thus, if any subspaces are to fail to split, they must be of "intermediate size" (infinite in both dimension and co-dimension). It is perhaps not even obvious that there exists an example of a subspace which fails to split. It turns out that such examples do exist, but that it is surprisingly difficult to display one. We now give such an example (an example we shall not use again, but which is nonetheless perhaps worthwhile, since it gives some insight into the nature

of the notion of splitting).
Examples. For purposes of this example, we define a functional on Banach space E as a linear mapping f from E to the reals such that $|f(x)| \leq |x|$ for every x in E.

Denote by E^1 the Banach space consisting of all sequences, $(r_1, r_2, \ldots)$, of real numbers the sum of whose absolute values, $\Sigma|r_i|$ is finite, where the norm of such a sequence is this sum. We next wish to claim that this Banach space E^1 has the following property: Given any sequence, $x_1, x_2, \ldots$, of unit vectors in E^1, there is some functional f on E^1 such that the sequence of numbers $f(x_1), f(x_2)\ldots$. fails to converge to zero. The proof is as follows. For any finite subset L of the positive integers, and any positive integer i, denote by S^i_L the real number obtained by taking the sum $\Sigma_L r_j$, where $x_i = (r_1, r_2, \ldots)$. If, for any finite L, we did not have $\lim_{i\to\infty} S^i_L = 0$, then we would be done: let f be the functional which assigns to $x = (s_1.s_2, \ldots)$ the number $f(x) = \Sigma_L s_j$. So, we may suppose that $\lim_{i\to\infty} S^i_L = 0$ for every L. Choose finite L^1 such that $|S^1_{L^1}| \geq 1/4$. Then, find integer n_1 such that $|S^j_{L^1}| \leq 1/16$ whenever $j \geq n_1$. Next, choose finite L^2, disjoint from L^1, such that $|S^{n_1}_{L^2}| \geq 1/4$.

Then, find integer $n_2 > n_1$ such that $|S^j_{L^2}| \leq 1/32$ whenever $j \geq n_2$. Next, choose finite L^3, disjoint from L^1, such that $|S^{n_2}_{L^3}| \geq 1/4$. Continuing in this way, we obtain disjoint finite sets $L^1, L^2, \ldots$. We next introduce a sequence $(a_1, a_2, \ldots)$ of real numbers according o the following rule: $a_i = 0$ if i is none of $L^1, L^2, \ldots$ $a_i = +1$ if i is in some L^j with $S^{n_{j-1}}_{L^j}$ positive, and $a_i = -1$ if i is in some L^j with $S^{n_{j-1}}_{L^j}$ negative. Now let f be the following functional on Banach space E^1, for $x = (r_1, r_2, \ldots)$ in E^1, $f(x) = r_1 a_1 + r_2 a_2, + \ldots$ (noting that this sum converges, since the sum of the absolute values of the r_i is finite, and the $|a_i|$ are bounded). But, by construction, $|f(x_i)| \geq 1/8$ for i one of the n_j. Hence, $f(x_1), f(x_2), \ldots$ cannot converge to zero.

We next denote by E^2 the Banach space consisting of all sequences $(r_1, r_2, \ldots)$ of real numbers for which the sum $\Sigma|r_i|^2$ is finite, where the norm of such a sequence is the square root of this sum. We claim that the Banach space E^2 fails to satisfy the property that we just showed for E^1. That is: there exists a sequence $x_1, x_2, \ldots$ of unit vectors in E^2 such that $f(x_i)$ goes to zero for every functional f. Indeed, let $x_1 = (1, 0, 0, \ldots)$, $x_2 = (0, 1, 0, 0, \ldots)$, $x_3 = (0, 0, 1, 0, \ldots)$, etc. Set $a_i = f(x_i)$, where f is some functional. then for any numbers $r_1, r_2, \ldots$ with $\Sigma|r_i|^2$ finite, the sequence in E whose n^{th} term is $y_n = r_1 x_1 + \ldots + r_n x_n$ is Cauchy, whence it converges to some vector y in E. Then, since f is a functional, the sequence $f(y_i)$ of real numbers must converge, whence the sum $\Sigma r_i a_i$ must converge. That is to say, the numbers a_i must be such that $\Sigma r_i a_i$ is finite whenever $\Sigma|r_i|^2$ is finite. But this is possible only if $\Sigma|a_i|^2$ is finite. But this, in turn, requires that the a_i converge to zero. We conclude: $f(x_i)$ approaches zero. But this is what we wanted to show.

Thus, we now have two Banach spaces which differ in a certain respect. [In particular, we know that there exists no isomorphism from E^1 to E^2.]

The next step is the introduction of a certain mapping from E^1 to E^2. Let $x_1, x_2, \ldots$ be the sequence (ordered in some way) of all vectors in E^2 of the following form: $a(r_1, r_2, \ldots, r_n, 0, 0, \ldots)$, where $r_1, \ldots, r_n$ are rational, and the number a is so chosen that these are unit vectors in E^2. The mapping ψ is now defined as follows. Given any element $x = (s_1, s_2, \ldots)$ in E^1, let $\psi(x)$ be that element of E^2 which is the limit y of the Cauchy sequence in E^2 whose n^{th} term is $y_n = s_1x_1 + \ldots + s_nx_n$ (noting that $y_1, y_2, \ldots$ is indeed a Cauchy sequence since $|x_i| = 1$ for every i, while $\Sigma|s_j|$ is finite). This mapping ψ is clearly linear. Furthermore, it is onto E^2 (for finite linear combinations of the x_i are dense in E^2). Furthermore, this mapping is even norm-decreasing (for, since the x_i are unit, $|s_1x_1 + \ldots + s_nx_n| \leq |s - 1| + \ldots + |s_n|$).

Denote by K the kernel of the mapping ψ, i.e., the set of all vectors x in E^1 with $\psi(x) = 0$. Then, since ψ is linear, K is a vector subspace of vector space E^1. Further, since ψ is norm-decreasing, K is a closed subset of E^1 (for, given sequence $z_1, z_2, \ldots$ in K, converging to z, then $|\psi(z) - \psi(z_i)| \leq |z - z_i|$, whence, since the z_i converge to z, the $\psi(z_i)$ converge to $\psi(z)$. But $\psi(z_i) = 0$ and so we must have $\psi(z) = 0$, i.e., x must also be in K.)

Thus, we now have Banach space E^1, and subspace K. We claim, finally, that this subspace of E^1 does not split. Suppose that it did, say with complementary subspace G: We obtain a contradiction. Since the mapping ψ is onto, the subspace G must be isomorphic with Banach space E^2. Choose a sequence of vectors in E^2, each unit, but such that for any functional f on E^2, f, applied to this sequence, approaches zero. Since E^2 is isomorphic with G, we obtain a similar sequence in E^1. The resulting sequence in $E^1, w_1, w_2, \ldots$ will have the property that, for every functional f on E^1, limit $f(w_i) = 0$ (since every functional on E^1 can also be regarded as a functional on the subspace G). Furthermore, the norms of the vectors $w_1, \ldots$ will be bounded below (since the w_i arose originally from a sequence of unit vectors in E^2). But, (after rescaling the w_i) so that they are unit) this contradicts the property of E^1 that we showed at the beginning: whenever $f(w_i) \to 0$ for every f, $|w_i| \to 0$. This contradiction establishes that the subspace K of E^1 has no complementary subspace, completing the example.

The subspace of converging sequences in the Banach space of bounded sequences also has no complementary subspace. Every subspace of E^2 has a complementary subspace.

The general rule seems to be that either one can discover with relative ease some explicit complementary subspace, or none exists. Note, incidentally, that, in the finite-dimensional case, every subspace splits. It is for this reason that splitting is not normally mentioned in the finite-dimensional case.

6. Bounded Linear Mappings

The "structure-preserving" (and, therefore, the useful, and interesting) mappings between Banach spaces are what are called bounded linear mappings. In this section, we introduce these, and give some examples and properties.

Let E and F be Banach spaces. A mapping T from E to F is called a *bounded linear mapping* provided i) T is a linear mapping of vector spaces (i.e., for every x and y in E, and every real number b, $T(x + by) = T(x) + bT(y)$), and ii)T is bounded (i.e., there exists a number a such that, for every x in E, $|T(x)| \leq a|x|$).

We have already used this notion, although not this term, several times. what was called a functional on page 18 was precisely a bounded linear mapping from E to $\mathbb{R}$. Banach spaces $E, |\ |$ and $E, \{\ \}$ are equivalent (page 6) if and only if both the identity mapping from $E, |\ |$ to $E, \{\ \}$ and its inverse are bounded linear mappings. the open mapping theorem can be restated thus: if a bounded linear mapping from one Banach space to another is one-to-one and onto (so its inverse exists), then that inverse is also a bounded linear mapping.

For T a bounded linear mapping from Banach space E to Banach space F, the smallest (necessarily non-negative) number a such that $T|(x)| \leq a|x|$ for every x is called the *norm* or *bound* of T, and is written $|T|$.

Example. Let E be the Banach space of bounded sequences (with norm the least upper bound of absolute values of entries), and F the Banach space of convergent sequences (same norm). The identity mapping from F to E (every element of F is automatically an element of E) is a bounded linear mapping. The mapping from F to $\mathbb{R}$ which assigns to each sequence its limit is a bounded linear mapping. The mapping from E to $\mathbb{R}$ which assigns to $(r_1, r_2, \ldots)$ the number r_{137} is a bounded linear mapping. Each of these mappings has norm one.

Example. Let E be the Banach space of sequences the sum of the absolute values of whose entries converges, and let F be the same, with the sum of the squares of absolute values. The identity mapping from E to F is a bounded linear mapping, with norm one. The mapping ψ in the beginning of page 19 is a bounded linear mapping: the remarks of that paragraph show that its norm less than or equal to one.

Example. Let E be the Banach space of C^n functions on the reals, the value of whose first n derivatives are bounded (page 5). Let F be the Banach space of C^{n-1} functions. Let T be the mapping "take the derivative" from E to F. Then T is a bounded linear mapping (with norm one). The mapping from E to R given by "evaluate the derivative at real number 7" is also a bounded linear mapping with norm one (note that in this case, however, there is no nonzero x with $|T(x)| = |T|\,|x|$).

In the finite-dimensional case, the second condition for a bounded linear mapping is unnecessary: every linear mapping from one finite-dimensional Banach space to another is necessarily bounded. It turns out that boundedness does not follow from linearity in the infinite-dimensional case, but examples seem to be rather difficult to display.

Example. Let each of E and F be the Banach space of bounded sequences of real numbers. Denote by A the subset of E consisting of sequences which are zeros after a while. Then A is a vector subspace of vector space E (but not a Banach subspace). For $(r_1, \ldots, r_n, 0, 0, \ldots)$ in A, set $T(r_1, \ldots, r_n, 0, \ldots = (r_1, 2r_2, 3r_3, \ldots nr_n, 0, \ldots)$. Clearly, if we can extend the action of this T linearly to all of E, we shall have our counterexample. Consider the collection of all vector subspace E having in common with A only the zero vector. Ordering by inclusion, we have that the hypothesis of Zorn's lemma is satisfied. Let B be a maximal element: Then, by maximality, every vector in E can be written in one and only one way as the sum of one vector in A and one in B. For $x = y + z$ in E, with y in A and z in B, set $T(x) = T(y)$. This is the desired extension.

Finally, we consider candidates for subspaces defined by a bounded linear mapping. Let E and F be Banach spaces, and let T be a bounded linear mapping from E to F. We denote by Ker T, the *kernel* of T, the set of all vectors x in E such that $T(x) = 0$. Then Ker T is clearly a vector subspace of vector space E. We claim, furthermore, that Ker T is closed. Indeed, let $x_1, x_2, \ldots$ be a sequence in Ker T, converging to x in E. Then, since $|T(x) - T(x_i)| \leq |T|\,|x - x_i|$, the sequence $T(x_1), T(x_2), \ldots$ in F converges to $T(x)$ in F. But the $T(x_i)$ are zero, whence $T(x) = 0$, whence x is also in Ker T. Thus, Ker T is a subspace of E. Next, denote by Im T, the *image* under T, the set of all vectors in F of the form $T(x)$ for some x in E. Then Im T is clearly a vector subspace of vector space F. However, Im T is not in general closed.

Example. Let E be the Banach space of bounded sequences, and F that of convergent sequences. Let T be the bounded linear mapping from E to F with action $T(r_1, r_2, \ldots) = (r_1, r_2/2, r_3/3, \ldots)$ (so $|T| = 1$). Then the vector $(1/\sqrt{1}, 1/\sqrt{2}, 1/\sqrt{3}, \ldots)$ in F is certainly in the closure of Im T, but is the image of no vector in E (for the only candidate is $\sqrt{1}$, $\sqrt{2}$, $\sqrt{3}, \ldots$), which is not in E. In the finite-dimensional case, Im T, as a vector subspace, is always closed.

We shall next be concerned with spaces of bounded linear mappings. Fix Banach spaces E and F. We denote by $\mathcal{L}(E;F)$ the set of all bounded linear mappings from E to F. We define addition and multiplication by reals within this set $\mathcal{L}(E;F)$ thus: For T and T' in $\mathcal{L}(E:F)$, and a a real number, let $T + aT'$ be that mapping with $(T + aT')(x) = T(x) + aT'(x)$. [Note that the linear mapping $T + aT'$, so defined, is indeed bounded, for $|(T + aT')(x)| = |T(x) + aT'(x)| \leq |T(x)| + |a|\ |T'(x)| \leq (|T| + |a|\ |T'|)\ |x|$. Thus, $|T + aT'| \leq |T| + |a|\ |T'|$.] With these operations, the set $\mathcal{L}(E;F)$ becomes a vector space. We next note that "take the norm" is a norm on the vector space $\mathcal{L}(E;F)$, the three properties for a norm being immediate from the little formula just derived. In fact, this vector space with norm, $\mathcal{L}(E;F)$ is even a Banach space, i.e., it is complete under its norm. Proof: Let $T_1, T_2, \ldots$ be a Cauchy sequence in $\mathcal{L}(E;F)$. Then for each vector x in E, we have, since $|T_i(x) - T_j(x)| \leq |T_i - T_j|\ |x|$, that $T_1(x), T_2(x), \ldots$ is a Cauchy sequence in F. Hence, this sequence converges to some vector in F, which we denote $T(x)$, thus defining a (clearly linear) mapping T from E to F. We next note that this T is bounded, for, for any x in E, $|T(x)| = \lim|T_i(x)| = \lim|T_1(x) - (T_1 - T_i)(x)| \leq |T_1(x)| + \lim|(T_1 - T_i)(x)| \leq (|T_1| + \lim|T_1 - T_i|)\ |x|$. We show that the T_i converge to T in $\mathcal{L}(E;F)$. Given positive ϵ, choose i so that $|T_i - T_j| \leq \epsilon$ for $j \geq i$. Then $|(T - T_i)(x)| \leq \underset{j\to\infty}{\text{limit}}|(T - T_j)(x) + (T_j - T_i)(x)| \leq \lim|(T - T_j)(x)| + \lim|(T_j - T_i)(x)| \leq \epsilon|x|$, whence $|T - T_i| \leq \epsilon$. Thus, we conclude that $\mathcal{L}(E;F)$ is a Banach space. This construction yields many new Banach spaces from those we have considered already.

One generalizes the notion of a bounded linear mapping to several independent variables as follows. Let $E_1, \ldots, E_n$ and F be Banach spaces. A mapping T which assigns to an n-tuple of vectors, one from each of the E_i, a vector in F is called a *bounded multilinear mapping* provided, i) T is multilinear, i.e., linear in each variable separately (e.g., $T(x_1, x_2 + ax'_2, x_3, \ldots, x_n) = T(x_1, x_2, \ldots, x_n) + aT(x_1, x'_2, \ldots x_n)$), and ii) T is bounded, i.e., there is a number a such that $|T(x_1, \ldots, x_n)| \leq a\ |x_1||x_2|\ldots|x_n|$. The smallest such number is called the *bound*, or *norm*, of T, and is again denoted $|T|$. The various properties of bounded linear mappings carry over almost immediately to similar properties of bounded multilinear mappings. In particular, the set of all such form $E_1, \ldots, E_n$ to F, denoted $\mathcal{L}(E_1, \ldots, E_n; F)$, is a Banach space.

There is a certain sense, which we now explain, in which multilinear mappings are extraneous. Let E, F, and G be Banach spaces. We define a mapping ψ from the Banach space $\mathcal{L}(E, F; G)$ to $\mathcal{L}(E; \mathcal{L}(F; G))$ as follows: For T in $\mathcal{L}(E, F; G)$, let $\psi(T)$ be that element of $\mathcal{L}(E; \mathcal{L}(F; G))$ which sends the vector x in E to that element of $\mathcal{L}(F; G)$ whose action on vector y in F is the vector $T(x, y)$ in G. This mapping is clearly linear. It is, in fact, invertible: Indeed, ψ^{-1} sends the element W of $\mathcal{L}(E; \mathcal{L}(F; G))$ to that element pf $\mathcal{L}(E, F; G)$ whose action on x, y (x in E,; y in F) is the element

$[W(x)](y)$ of G (nothing that $W(x)$ is an element of $\mathcal{L}(F;G)$). It is furthermore immediate from the definitions (although confusing in detail) that ψ is norm-preserving. Thus, the Banach spaces $\mathcal{L}(E,F;G)$ and $\mathcal{L}(E;\mathcal{L}(F;G))$ are equivalent. Those who like $\mathcal{L}$'s will note that our ψ is a preferred element of $\mathcal{Z}(\mathcal{L}(E,F;G);\mathcal{L}(E;\mathcal{L}(F;G))$. Spaces of multilinear mappings can thus always be expressed in terms of iterated $\mathcal{L}$'s.

We conclude this section with the introduction of two additional "preferred objects". Let E and F be Banach spaces. Denote by α the following element of $\mathcal{L}(E,\mathcal{L}(E;F);F)$: For x in E and T in $\mathcal{L}(E;F)$, $\alpha(x,T) = T(x)$ (an element of F). Thus, α is the multilinear mapping whose action is "action of an element of $\mathcal{L}(E;F)$ on an element of E". Multilinearity of α is obvious; we have only to show roundedness. But we have $|\alpha(x,T)| = |T(x)| \leq |T|\ |x|$. Thus, not only is α bounded, but its norm is one. For the second object, let E, F and G be Banach spaces. Let β denote the following element of $\mathcal{L}(\mathcal{L}(E;F),\ \mathcal{L}(F;G);\ \mathcal{L}(E;G))$: For T in $\mathcal{L}(E;F)$ and U in $\mathcal{L}(F;G)$, let $\beta(T,U) = U \cdot T$ (composition of mappings: a mapping from E to G). We must first show that the linear mapping $U \cdot T$ is actually in $\mathcal{L}(E;G)$, i.e., that it is bounded. For x in E, $|(U \cdot T)(x)| = |U|\,[T(x)]| \leq |U|\ |T(x)| \leq |U|\ |T|\,|x|$, whence $|U| \cdot |T| \leq |U|\ |T|$. Thus, $U \cdot T$ is indeed in $\mathcal{L}(E;G)$, and so our mapping β is indeed well-defined. It is obviously that this β is multilinear. Thus, we have only to show that β is bounded. But for T in $\mathcal{L}(E;F)$ and U in $\mathcal{L}(F;G)$, we have $|\beta(T,U)| = |U \cdot T| = |U \cdot T| \leq |U|\ |T|$. Thus, is indeed bounded, and in fact we have $|\beta| = 1$. This β, of course, just performs for us the operation of "composition of bounded linear mappings". That β is itself a bounded multilinear mapping expresses the basic properties of composition.

What is so nice about all this is that everything in sight is a Banach space: Everything one tries to do with Banach spaces yields just other Banach spaces. The result is that one has to learn in detail but a single kind of mathematical object.

7. Derivatives

Elementary differential calculus (of several variables) is of course carried out in $\mathbb{R}^n$. But $\mathbb{R}^n$ is a particular example of a Banach space. In this section, we shall see that finite-dimensionality plays essentially no role in differential calculus: We shall, indeed, repeat the basic ideas of this subject for Banach spaces.

Fix Banach spaces E and F, Fix an open subset U of E (so, for every point x in U, there is a ball centered at x in U). Let f be a mapping from the set U to the set F. [In the finite-dimensional case, with E n–dimensional and F m–dimensional, f is represented as m functions of n real variable.] [Note that f is only defined on U. whenever we speak of the action of f on a ball centered at a point of U, we shall suppose implicitly that this ball is small enough so that it is within U.]

Fix a point x_0 of U. The mapping f will be said to be *continuous* at x_0 provided that, for every positive number ϵ, there is a positive number δ such that $|f(x) - f(x_0)| \leq \epsilon$, whenever $|x - x_0| \leq \delta$. This f is said to be just *continuous*, or C^0, if it is continuous at every point of U.

We next wish to get hold of the "rate of change of $f(x)$ with x". We first introduce the notion of "zero rate of change". The mapping f is said to be *tangent* at x_0 provided that, for every positive number ϵ, there is a positive number δ such that $|f(x) - f(x_0)| \leq \epsilon|x - x_0|$ whenever $|x - x_0| \leq \delta$. Thus, f is tangent at x_0 provided "the deviation of $f(x)$ from $f(x_0)$ relative to the deviation of x from x_0 becomes as small as we wish (ϵ) whenever x is sufficiently close (δ) to x_0." Note that, if f is tangent at x_0, then f is automatically continuous at x_0 (although the converse, as we shall see shortly, is false).

The mapping f is said to be *differentiable* at x_0 provided that there exists a bounded linear mapping T from E to F such that $f(x) - f(x_0) - T(x - x_0)$ is tangent at x_0. In other words, we require that, given any positive ϵ, there is a positive δ such that $|f(x) - f(x_0) - T(x - x_0)|/|x - x_0| \leq \epsilon$ whenever $x \neq x_0$ and $|x - x_0| \leq \delta$. Thus, for differentiability we require that "$f - f(x_0)$ can be approximated, up to tangency, by a bounded linear mapping". This T, called the *derivative* of f at x_0, is written $Df(x_0)$, so $Df(x_0)$ is an element of $\mathcal{L}(E; F)$. [This is what we expect. In the finite-dimensional case, $Df(x_0)$ is represented by n partial derivatives of the m functions representing f. But this $n \times m$ matrix can be regarded as a linear mapping from n–dimensional E to m–dimensional F.] We shall see shortly that the derivative is unique if it exists. If f is differentiable at every point x_0 of U, then f is said to be just *differentiable*. In this case Df assigns, to each point x_0 of U, an element $Df(x_0)$ of $\mathcal{L}(E; F)$. Thus, Df is a mapping from U (an open subset of Banach space E) to $\mathcal{L}(E; F)$ (a Banach space). If this mapping Df is continuous (examples in finite dimensions show that it need not always be), then f is said to be *continuously differentiable*, or C^1.

In the finite-dimensional case, continuous, differentiable, and continuously differentiable reduce, of course, to their usual meanings.

Example. Let E and F be Banach spaces, and fix bounded linear mapping T from E to F. Set $U = E$, and let f be the mapping from U to F with $f(x) = T(x)$. Then, for any given x_0 in U, we have for all x $|f(x) - f(x_0)| \leq |T|\,|x - x_0|$. Thus, f is continuous (given ϵ, choose $\delta = \epsilon/|T|$). By definition of the norm, we have that, for fixed x_0 in U and for any number $a > 1$, there is an x in U such that $|f(x) - f(x_0)| \geq a|T|\,|x - x_0|$. But by linearity this inequality continues to hold if x is replaced by any vector on the line joining x and x_0. Hence, f cannot be tangent at x_0 unless f is the zero linear mapping (a mapping which is necessarily tangent at every x_0). We next verify that f is differentiable at x_0. Indeed, $f(x) - f(x_0) - T(x - x_0) = 0$ for all x, whence the left side is certainly tangent at x_0. Hence, $Df(x_0) = T$. Thus, Df assigns to the element x_0 of U the fixed element T of $\mathcal{L}(E; F)$. Since this (constant) mapping Df from U to $\mathcal{L}(E; F)$ is certainly continuous, our f is C^1.

Example. Let f and f' both be continuous. Then so is $f + f'$ (with action $(f + f')(x) = f(x) + f'(x)$). Indeed, given ϵ, find δ_1 such that $|f(x) - f(x_0)| \leq \epsilon/2$ whenever $|x - x_0| \leq \delta_1$, and δ_2 such that $|f'(x) - f'(x_0)| \leq \epsilon/2$ whenever $|x - x_0| \leq \delta_2$. Then, whenever $|x - x_0| \leq \min(\delta_1, \delta_2)$, we have $|(f + f')(x) - (f + f')(x_0)| \leq \epsilon$. By a similar argument, the sum of two functions tangent at x_0 is again tangent at x_0. Next, suppose that each of f and f' is differentiable at x_0. Then, since the functions with action $f(x) - f(x_0) - Df(x_)(x - x_0)$ and $f'(x) - f'(x_0) - Df'(x_0)(x - x_0)$ are both tangent at x_0, so is their sum, $(f + f')(x) - (f + f')(x_0) - (Df(x_0) + Df'(x_0))(x - x_0)$. Hence, $f + f'$ is also differentiable at x_0, and, furthermore, $D(f + f')(x_0) = Df(x_0) + Df'(x_0)$. But now, since the sum of continuous functions is continuous, it is immediate from $D(f + f') = Df + Df'$ that the sum of two C^1 functions is C^1.

It follows from these two example that the derivative $Df(x_0)$, if it exists, is unique. Indeed, let T and T' be two (so each is in $\mathcal{L}(E; F)$). Then both $f(x) - f(x) - T(x - x_0)$ and $f(x) - f(x_0) - T'(x - x_0)$ are tangent at x_0, whence their difference, the mapping $T - T'$, is also tangent. But the only bounded linear mapping which is tangent at x_0 is the zero one, whence $T = T'$.

Example. Let each of E and F be the Banach space of continuous, bounded, real-valued functions on the reals. Let $U = E$, and let f be the mapping from U to F which function φ to φ^2 (noting that the square of a continuous bounded function is another). Fix element φ_0 of U. Then f continuous at φ_0. Indeed, $|f(\varphi) - f(\varphi_0)| = |\varphi^2 - \varphi_0^2| = |(\varphi - \varphi_0 + 2\varphi_0)(\varphi - \varphi_0)| \le (|\varphi - \varphi_0| + 2|\varphi_0|)\,|\varphi - \varphi_0|$. Hence, by making $|\varphi - \varphi_0|$ sufficiently small we also make $|f(\varphi) - f(\varphi_0)|$ as small as we wish. We next show that this f is in fact differentiable at φ_0. As a general rule, it is difficult ro show differentiability without first making a guess as to what the derivative is to be. Taking our cue from elementary calculus, we guess as follows: Let α be the bounded linear mapping from E to F which sends element φ of E to element $\alpha(\varphi) = 2\varphi\varphi_0$ of F. Then we have $f(\varphi) - f(\varphi_0) - \alpha(\varphi - \varphi_0) = \varphi^2 - {\varphi_0}^2 - 2\varphi_0(\varphi - \varphi_0) = (\varphi - \varphi_0)^2$. Thus, $|f(\varphi) - f(\varphi_0) - \alpha(\varphi - \varphi_0)| \le |\varphi - \varphi_0|^2$. Clearly, then, (choose $\delta = \epsilon$), the mapping on the left is tangent at φ_0. We conclude that f is differentiable, and $Df(\varphi_0) = \alpha$. Finally, we claim that this mapping Df from U to $\mathcal{L}(E;F)$ is itself continuous. Indeed, $|Df(\varphi) - Df(\varphi')| = |Df(\varphi - \varphi')| \le 2|\varphi - \varphi'|$. Thus, the mapping f is C^1.

Example. Let E be any Banach space, and let F be the Banach space of reals. Set $U = E$, and let f be the mapping from U to F which sends x in U to $f(x) = |x|$. Then, given x_0 in U, f is continuous at x_0, for $|f(x) - f(x_0)| = ||x| - |x_0|| \le |x - x_0|$ [Choose $\delta = \epsilon$.]. However, f need not necessarily be differentiable at x_0, even for $x_0 \neq 0$. For example, let E be $\mathbb{R}^2$, with norm $|(r, r')| = \max(|r|, |r'|)$. Then f is not differentiable, e.g., at $x_0 = (1, 1)$.

Let E and F be Banach spaces, U an open subset of E, and f a C^1 mapping from U to F. Then Df is a continuous mapping from U to $\mathcal{L}(E;F)$. But this Df is a mapping from an open subset (U) of a Banach space to a Banach space. If this mapping is in fact differentiable, and if its derivative, DDf is continuous, then f is said to be twice continuously differentiable, or C^2. In this case, DDf is a continuous mapping from U to $\mathcal{L}(E;\mathcal{L}(E;F))$. Similarly, if this mapping is C^1 (so $DDDf$ exists as a continuous mapping from U to $\mathcal{L}(E;\mathcal{L}(E;\mathcal{L}(E;:F))))$, then f is said to be C^3. Similarly for C^p, p a non-negative integer. Finally, f is said to be C^∞ if f is C^p for every p.

Example. The mapping of the first example on page 26 is C^∞. Since Df is constant ($Df(x_0) = T$ for every x_0), $DDf = 0$ $DDDf = 0$, etc.

Example. The sum of two C^p mappings is C^p. This is immediate from the facts that the sum of two C^1 mappings is C^1, that the derivative of the sum is the sum of the derivatives.

Example. The mapping of the first example on page 26 (beginning on the previous page) is C^∞. Indeed, DDf is that element of DDf is that element of $\mathcal{L}(E;\mathcal{L}(E;P))$ which sends φ in E to the mapping from E to P which sends φ' to the element $2\varphi\varphi'$ of F. Hence, DDf is constant. So, $DDDf = 0$, etc. [Just what one expects of a "quadratic mapping".]

Example. Let E and F be Banach spaces, and T an element of $\mathcal{L}(E, \ldots, E; F)$. Set $f(x) = T(x, \ldots, x)$. Then f is C^∞.

We next show that composition of C^p mapping is C^p.

Theorem. Let E, F and G be Banach spaces, U an open subset of E and V an open subset of F. Let f be a C^p mapping from U to F with $f(U) \subset V$ and g a C^p mapping from V to G. Then the mapping $g \cdot f$ from U to G is also C^p.

Proof: suppose first that f and g are C^0. Then, given x_0 in U and positive ϵ, first choose δ' such that $|g(y) - g(f(x_0))| \le \epsilon$ whenever $|x - x_0| \le \delta'$. Then choose δ, such that $|f(x) - f(x_0)| \le \delta'$ whenever $|x - x_0| \le \delta$. Then, for $|x - x_0| \le \delta$, we have $|g \cdot f(x) - g \cdot f(X_0)| \le \epsilon$. That is, $g \cdot f$ is C^0.

Next, suppose that f and g are C^1, Fix x_0 in U. Then $Dg(f(x_0))$ is an element of $\mathcal{L}(F; G)$, while $Df(x_0)$ is an element of $\mathcal{L}(E; F)$. Their composition is thus an element of $\mathcal{L}(E; G)$. Note first that the mapping with action $f(x) - f(x_0) - Df(x_0)(x - x_0)$ is tangent at $x = x_0$. Applying $Dg(f(x_0))$ to this expression, using the fact that the composition of a bounded linear mapping with a function tangent is a function tangent, we have that $Dg(f(x_0))(f(x) - f(x_0)) - Dg(f(x_0))Df(x_0)(x - x_0)$ is tangent at $x = x_0$. But, since $g(y) - g(f(x_0)) - Dg(f(x_0))(y - f(x_0))$ is tangent at $y = f(x_0)$, and since f is continuous, $g(f(x)) - g(f(x_0)) - Dg(f(x_0))(f(x) - f(x_0))$ is tangent at $x = x_0$. Adding these two, we obtain: $g(f(x)) - g(f(x_0)) - Dg(f(x_0))Df(x_0)(x - x_0)$ is tangent at $x = x_0$. But this is precisely the statement that $g \cdot f$ is differentiable at x_0, with $D(g \cdot f)(x_0) = Dg(f(x_0))Df(x_0)$. Next note that $f(x)$ is continuous in x and $Dg(y)$ is continuous in y, whence their composition, $Dg(f(x))$ is continuous in x. Further, $Df(x)$ is continuous in x. But the element of $\mathcal{L}(\mathcal{L}(E; F), \mathcal{L}(F; G); \mathcal{L}(E; G))$ which composes is also continuous (pp 24). Hence, $Dg(f(x))Df(x)$ is a continuous mapping from U to $\mathcal{L}(E; G)$. We conclude that $g \cdot f$ is C^1.

Now suppose that both f and g are C^2. Since $f(x)$ is C^1 in x, and $Dg(y)$ is C^1 in y, $Dg(f(x))$ is C^1 in x (last paragraph). Also, $Df(x)$ is C^1 in x. But composition is also C^1 (since it is bilinear). Hence, $Dg(f(x))\, Df(x)$ is C^1 in x. That is, $D(g \cdot f)$ is C^1, whence $g \cdot f$ is C^2. Continue in the obvious way to C^p.

We next show that operation "taking the inverse" is also C^∞.

Example. Let E and F be Banach spaces. Denote by U the subset of $\mathcal{L}(E; F)$ consisting of all T therein which are invertible. We first show that this subset U is open. Fix T_0 in U. Then T_0 is an isomorphism from E to F. Consider the isomorphism from $\mathcal{L}(E; F)$ to $\mathcal{L}(E; E)$ which sends T in $\mathcal{L}(E; F)$ to $T\, T_0^{-1}$. Under this isomorphism, T_0 itself is sent to I, the identity on E. It suffices, therefore, to show that there is some ball centered at I in $\mathcal{L}(E; E)$ every element of which is invertible. Let B be the ball with radius $1/2$. Let W be an element of B. Consider the sequence $V - 1 = I, V - 2 = I + (I - W)$, $V_3 = I + (I - W) + (I - W)(I - W)$, etc. This is a Cauchy sequence in $\mathcal{L}(E; E)$, for $|V_{i+1} - V_i| = |(I - W)^i| \le |I - W|^i \le 1/2^i$. Hence, this sequence converges to some element V of $\mathcal{L}(E; E)$. We now have, by direct computation, $WV_i - I = V_i - V_{i+1}$. Taking the limit of each side as i increases, noting that that on the right is zero, we obtain $WV = I$, and, similarly, $VW = I$. Thus, V is the inverse of W, and so W is invertible. We have shown so far that our subset U of $\mathcal{L}(E;F)$ is open. Note also that, above, $|W^{-1}| \leq 1 + 1/2 + 1/4 + \ldots = 2$.

Next, introduce the mapping f from U to $\mathcal{L}(F;E)$ with the following action: For T in U, $f(T) = T^{-1}$. We next show that this mapping f is continuous. Fixing T_0 in U, we have for T in U, $f(T) - f(T_0) = T^{-1} - {T_0}^{-1} = -T^{-1}(T - T_0){T_0}^{-1}$, whence $|f(T) - f(T_0)| \leq |T^{-1}||T - T_0||{T_0}^{-1}|$. Given positive ϵ, choose δ such that, whenever $|T - T_0| \leq \delta$, $|T^{-1}| \leq M$, some constant (possible, from the observation above that, whenever $|I - W| \leq 1/2$, $|W^{-1}| \leq 2$), and such that $\delta \leq \epsilon / M{T_0}^{-1}$. Then, for $|T - T_0| \leq \delta$, we have $|f(T) - f(T_0)| \leq \epsilon$. Thus, f is continuous at T_0, which is arbitrary in U, and so f is continuous.

We next show that this mapping f is in fact differentiable at $T = T_0$. Its derivative at T_0 must be an element of $\mathcal{L}(\mathcal{L}(E;F);\mathcal{L}(F;E))$. From elementary calculus, we guess the derivative as follows: Let α be the bounded linear mapping from $\mathcal{L}(E;F)$ to $\mathcal{L}(F;E)$ which sends T in $\mathcal{L}(E;F)$ to $\alpha(T) = -{T_0}^{-1}T{T_0}^{-1}$. Then $f(T) - f(T_0) - \alpha(T - T_0) = T^{-1}_{-1} - {T_0}^{-1} + {T_0}^{-1}(T - T_0){T_0}^{-1} = T^{-1}(T - T_0)T_0(T - T_0){T_0}^{-1}$. Hence, we have $|f(T) - f(T_0) - \alpha(T - T_0)| \leq |T^{-1}||T_0|^{-1/2}|T - T_0|^2$. Now, given positive ϵ, choose δ such that, whenever $|T - T_0| \leq \delta$, we have $|T^{-1}| \leq M$ (some constant), and such that $\delta \leq \epsilon / M|{T_0}^{-1}|^2$. Then, for $|T - T_0| \leq \delta$, we have $|f(T) - f(T_0) - \alpha(T - T_0)| \leq \epsilon|T - T_0|$. We conclude that the mapping f from U to $\mathcal{L}(F;E)$ is indeed differentiable at T_0 in U, and that its derivative is α.

For each T_0 in U, we obtain, as above, the derivative of f at T_0, $Df(T_0)$. Since "take the inverse" and "compose" are continuous operations, this Df is a continuous mapping from U to $\mathcal{L}(F;E)$. So, f is C^1. But now, since "take the inverse" and "compose" are C^1 operations, Df is C^1, whence f is C^2. etc. So f is C^p.

All of the little calculations above are identical to those one is familiar with in the finite-dimensional case.

Finally, we establish the result in the present context which generalizes the finite-dimensional statement that "mixed partials commute". For this purpose, we note that DDf maps U to $\mathcal{L}(E;\mathcal{L}(E;F))$, and that $(E;F))$ is isomorphic with the Banach space $\mathcal{L}(E,E;F)$. Hence, we may regard $DDf(x_0)$ as an element of $\mathcal{L}(E,E;F)$.

Theorem. Let E and F be Banach spaces, U open is E, and f a C^2 mapping from U to F. Then DDf is symmetric: For x_0 in U and x, y in E, $DDf(x_0)(x,y) = DDf(x_0)(y,x)$.

Proof : Fix an element α of $\mathcal{L}(F;\mathbb{R})$. Let h be the function of two real variables a and b with action $h(a,b) = \alpha(f(x_0 + ax + by))$. [Thus, h is only defined for a and b small enough that $x_0 + ax = by$ is in U.] Using the chain rule (preceding theorem), we have $\partial h / \partial a = \alpha(Df(x_0 + ax + by)(x))$, and thus $\partial^2 h/\partial a\,\partial b = \alpha(DDf(x_0 + ax + by)(y, x))$. Evaluating at $a = b = 0$, we have $\partial^2 h/\partial a\,\partial b|_{a=b=0} = \alpha(DDf(x_0)(y, x))$, and similarly $\partial^2 h/\partial b\,\partial a|_{a=b=0} = \alpha(DDf(x_0)(x, y))$. But the left sides are equal (equality of mixed partials for real functions of two real variables), and so $\alpha(DDf(x_0)(y, x) - DDf(x_0)(x, y) = 0$ for every α. Thus, we shall be done if we can show that the only element z of F such that $\alpha(z) = 0$ for every α in $\mathcal{L}(F; \mathbb{R})$ is $z = 0$. Given z in F, choose complementary subspace G in F to the subspace spanned by z. Then any vector u in F can be written uniquely in the form $u = az + v$ with v in G. Let α have action $\alpha(u) = a$. Then α is bounded linear mapping (obviously linear, and bounded since α is bounded on both G and the subspace spanned by z, and since F is isomorphic with the product of these, by the open mapping theorem). but $\alpha(z) = 1$. So, this α is the thing we wanted to find. This completes the proof.

The proof above is rather tacky, because it uses itself in the finite-dimensional case, which is taken as known. Unfortunately, the proof of this theorem in finite-dimensions does not seem to be directly generalizable to infinite dimensions. I am aware of no more direct proof of the theorem above.

It follows immediately by repeated application of this result that higher mixed partials are also symmetric. For example, $DDDf(x_0)$, an element of $\mathcal{L}(E; \mathcal{L}(E; \mathcal{L}(E; F)))$, can also be regarded as an element of $\mathcal{L}(E, E, E; F)$. In this case we have, for x, y, and z in E and for fC^3, that $DDDf(x_0)(x, y, z) = DDDf(x_0)(y, x, z) = DDDf(x_0)(z, y, x)$, etc.

8. Mean Value Theorem; Inverse Mapping Theorem

We complete our treatment of elementary calculus in infinite dimensions by proving the two titled theorems.

The mean value theorem states that the derivative of a function, a measure of its rate of change, in fact bounds the actual changes in the value of the function.

Theorem (mean value). Let E and F be Banach spaces, U an open subset of E, and f a C^1 mapping from U to F. Let x and y be points of U such that the line segment γ joining x and y lies within U. Then $|f(y) - f(x)| \leq \underset{z \text{ in } \gamma}{\text{lub}} |Df(x)||y - x|$.

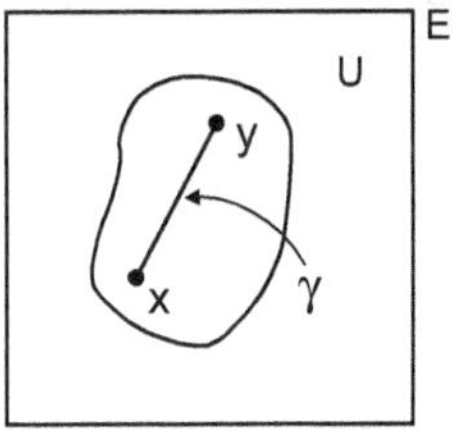

Proof: [Of course, γ is the set of vectors of the form $ax + (1 - a)y$, with $0 \leq a \leq 1$.] Fix, once and for all, positive number ϵ. Given z_0 on γ, the function with action $f(z) - f(z_0) - Df(z_0)(z - z_0)$ is tangent at $z = z_0$. Hence, there is a positive δ such that, whenever $|z - z_0| \leq \delta$, we have $|f(z) - f(z_0)| \leq (|Df(z_0)| + \epsilon)|z - z_0|$. Thus, for every z_0 on γ we have such a δ. Since γ is compact, we can find points $z_1, \ldots, z_n$ on γ such that $z_1 = x, z_n = y, z_i \leq z_{i+1}$, (in the obvious sense: 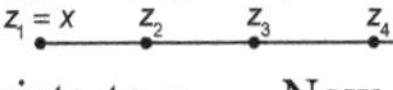

), and such that $|z_i - z_{i+1}|$ is less than the δ appropriate to z_{i+1}. Now, we have $|f(y) - f(x)| \leq |f(z_n) - f(z_1)|\ |f(z_n) - f(z_{n-1})| + \ldots + |f(z_2) - f(z_1)|$. But $|f(z_{i+1}) - f(z_i)| \leq |(Df(z_i)| = \epsilon)|z_{i+1} - z_i| \leq (\text{lub}\,|Df(z)| + \epsilon)|z_{i+1} - z_i|$. Hence, $|f(y) - f(x)| \leq (\text{lub}\,|Df(z)| + \epsilon)(|z_n - z_{n-1}| + \ldots + |z_2 - z_1|)$. But the sum on the right is just $|z_n - z_1| = |y - x|$. So, we have $|f(y) - f(x)| \leq (\text{lub}\,|Df(z)| + \epsilon)|y - x|$. Since, ϵ is arbitrary, the result follows.

One might imagine that another version of the mean value theorem, which on the one hand would be stronger and on the other would be more closely analogous to the one-dimensional mean value theorem, might be true: Under the conditions of the theorem above, there is a point z_0 of γ such that $f(y) - f(x) = Df(z)(y - x)$. This, however, is false.

Example. Let E be the reals, and $F = R^3$. Then a C^1 mapping f from $U = E$

to F represents a curve in $\mathbb{R}^3$. $Df(z)$ is the tangent vector to this curve at z in R.

On the other hand, $f(y)-f(x)$, is the vector joining these two points in $\mathbb{R}^3$. Thus, we must find a curve joining two points in $\mathbb{R}^3$, whose tangent vector is never parallel to the vector joining those points (for then we shall never have $f(y) - f(x) = Df(z)(y - x)$). But a spiral, as in the figure above, does the job.

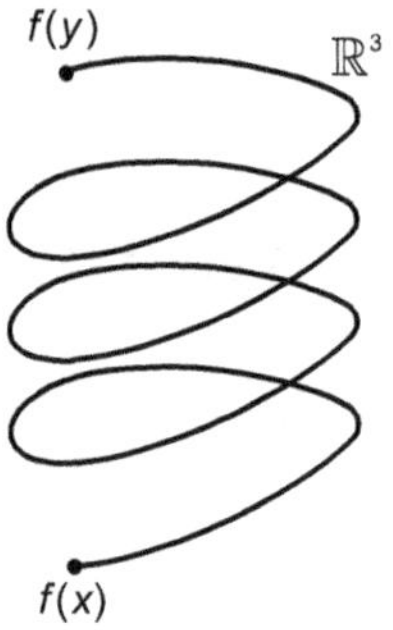

Example. Let $Df = 0$. Then, by the mean value theorem, f is constant (i.e., $f(x) = f(y)$ for all x, y such that the line segment joining them lies in U). Similarly, if $DDf = 0$, then the action of f is $f(x) = T(x) + x_0$, for some x_0 in E and T in $\mathcal{L}(E; F)$ (for Df is some constant T, and then $D(f - T(x)) = 0$, whence this is some constant x_0.)

The inverse function theorem states that, if a mapping is "invertible to first order some point, then it is actually invertible in some ball about that point". As a prerequisite to this theorem, we need a little fact about mappings (a fact which, indeed, is the basis for all existence and most uniqueness theorems about differential equations). Let C be a closed subset of a Banach space, and f a mapping from C to C such that, for some number $a < 1$, $|f(x) - f(y)| \leq a|x - y|$ for every x, y in C. Then there is one and only one x in C with $f(x) = x$ Proof: First note that, if there were two, x and y, then we would have $|y - x| = |f(y) - f(x)| \leq a|y - x|$, whence, since $a < 1, |y - x| = 0$, so $y = x$. Thus, we have only to show that $f(x) = x$ for some x. Choose y in C, and let $y_1 = y, y_2 = f(y_1), y_3 = f(y_2)$, etc. Then, since $|y_{i+1}-y_i| = |f(y_i)-f(y_{i-1})| \leq a|y_i-y_{i-1}|$, we have $|y_i-y_j| \leq a^j+\ldots+a^{j-1}$ (for $j > i$), whence the y_i form a Cauchy sequence. Since the y_i are in a Banach space, this sequence converges to some vector x; since C is closed and the y_i are in C, x is in C. But now $f(y_i) - x = y_{i+1} - x$. Taking the limit of each side as x increases, noting that that on the right is zero, we obtain $f(x) = x$.

Theorem (inverse mapping). Let E and F be Banach spaces, U open in E. and f a C^p mapping from U to F, with $p \geq 1$. Let, for some x_0 in U. the element $Df(x_0)$ of $\mathcal{L}(E; F)$ be invertible. Then there is some open subset V of U containing x_0 such that i) $f[V]$ is open in F, ii) f is one-to-one on V, and iii) the inverse mapping f from $f[V]$ to E is also C^p.

Proof: Let α be the mapping from F to E with action $\alpha(y) = (Df(x_0)^{-1}(y - f(x_0)) + x_0$. Then α is C^∞. Furthermore, the C^∞ β from E to F with action $\beta(x) = Df(x_0)(x - x_0) + f(x_0)$ is its inverse, i.e., $\alpha \cdot \beta$ is the identity on E and $\beta \cdot \alpha$ is the identity on F. We may consider, instead of f, $\alpha \cdot f$. that is to say, we may assume that f maps U to E, that $x_0 = 0$, that $f(x_0) = 0$, and that $Df(0)$ is the identity element of $\mathcal{L}(E; E)$.

Let g be the mapping from U to E which sends x in U to $g(x) = x - f(x)$.

Then $g(0) = 0$, and $Dg(0) = D$ (identity) $(0) - Df(0) = 0$. Since g is C^1, there is a positive ϵ such that, whenever $|x| < 2\epsilon$, $|Dg(x) < 1/2$. Then, from the mean value theorem, we have, for $|x| < 2\epsilon$, $|g(x)| < 1/2\ |x|$.

Next, fix y in E with $|y| \leq \epsilon/2$. Denote by g_y the mapping from U to E with action $g_y(x) = y+x-f(x)$. Then, for $|x| \leq \epsilon$ we have $|g_y(x)| = |y+g(x)| \leq |y| + |g(x)| \leq 1/2\epsilon + 1/2|x| \leq \epsilon$. Thus, denoting by B the set of all vectors x with $|x| \leq \epsilon$, we have that g_y is a mapping from B to B. Furthermore, for x and x' in B, we have $|g_y(x) - g_y(x')| = |g(x) - g(x')| \leq 1/2\ |x - x'|$, again by the mean value theorem (and the fact that $|Dg(x)| < 1/2$ for $|x| < 2$). Thus, g_y, as a mapping from the closed subset B of E to itself, a mapping with the contraction property of the previous page. Hence, there is one and only one x in B with $g_y(x) = x$, i.e., with $y + x - f(x) = x$, i.e., with $y = f(x)$.

Now let V consist of all x in U with $|x| < \epsilon$ and $|f(x)| < \epsilon/2$. Then V (as the intersection of U, an open ball, and the inverse image by f of an open ball) is open. Furthermore, $f[V]$ is just the set of y with $|y| < \epsilon/2$, and so is open in E. In addition, f is one-to-one on V (for, as we showed above, given y with $|y| \leq \epsilon/2$, there is one and only one x with $|x| \leq \epsilon$ and $f(x) = y$)

Denote by $\hat{f}$ the inverse of f, defined on $W = f[V]$. Then, for x and x' in V, $|x - x'| = |f(x) + g(x) - f(x') - g(x')| \leq |f(x) - f(x')| + |g(x) - g(x')| \leq |f(x) - f(x')| + 1/2|x - x'|$, whence $|x - x'| \leq 2|f(x) - f(x')|$. Setting $y = f(x)$ and $y' = f(x')$, this becomes $|\hat{f}(y) - \hat{f}(y')| \leq 2|y - y'|$. That is to say, this f is continuous.

With x, x', y and y' as above, we have $|\hat{f}(y) - \hat{f}(y') - (Df(x'))^{-1}(y - y')| = |(Df(x'))^{-1}(Df(x')(x - x') - f(x) + f(x'))| \leq |(Df(x'))^{-1}|\,|Df(x')(x - x') - f(x) + f(x')|$. Fixing x', the right side, regarded as a function of x, is tangent at x', since $\hat{f}$ is differentiable. Hence, by continuity of $\hat{f}$, the left side is tangent at $y = y'$. We conclude, therefore, that $\hat{f}$ is differentiable, and that $D\hat{f}(y) = (Df(\hat{f}(y))^{-1}$. Since $\hat{f}$ is continuous, Df is continuous, composition is continuous, and "take the inverse" is continuous, $D\hat{f}$ is continuous. That is, $\hat{f}$ is C^1.

If f is C^2, then, since $\hat{f}$ is C^1, Df is C^1, composition is C^1, and "take the inverse" is C^1, $D\hat{f}$ is C^1, and so $\hat{f}$ is C^2. Continuing in this way, if f is C^p, then so is $\hat{f}$.

There is a corollary of the inverse function theorem, called the implicit function theorem. It states that one can, under certain circumstances, "solve" equations of the form $f(x, y) = 0$ for x as a function of y. More precisely, let E and F be Banach spaces, and let U and V be open subsets of E and F, respectively. For each y in V, let f_y be a mapping from U to G. [We may thus regard f as a mapping from $U \times V$ to G; above, we write $f(x, y)$ for $f_y(x)$.] Suppose that i) $_y(x)$ is continuous on $U \times V$, ii) for each y in V, f_y is C^1, and $Df_y(x)$ is continuous on $U \times V$, and iii) for some x_0 in U and y_0 in V, we have $f_{y_0}(x_0) = 0$ and $Df_{y_0}(x_0)$, an element of $\mathcal{L}(E; G)$, is invertible. Then there is an open subset V_0 of V containing y_0, and a mapping

g from V_0 to U, such that $f_y(g(y)) = 0$ for every y in V_0. [Thus, $x = g(y)$ is the "solution" of $f_y(x) = 0$.] Proof: Set $h_y(x) = f_y(x) - f_y(x_0)$. Then $h_y(x_0) = 0$ for every y. Furthermore, $Dh_y = Df_y$. Hence, since $Df_{y_0}(x_0)$ is invertible, since $Df_y(x)$ is continuous, and since the invertible elements of $\mathcal{L}(E; G)$ form an open subset of $\mathcal{L}(E; G)$, we have that $Dh_y(x{-}0)$ is invertible for all y sufficiently close to y_0. Applying the inverse function theorem to h_y, for each y, we have: There is an open subset W of G containing 0, and a mapping s_y from W to E, for y sufficiently close to y_0, such that $h_y \cdot s_y$ is the identity on W. [We may choose W independent of y because the size of the "V" in the inverse function theorem is dictated by the size of "Df" in that theorem, and here $Df_y(x)$ is continuous.] Since $f_y(x)$ is continuous in y and x, it follows that, for y sufficiently close to y_0, $f_y(x_0)$ is in this W. Now set $g(y) = s_y(f_y(x_0))$. Then, since $h_y \cdots_y =$ identity, $h_y(g(y)) = f_y(x_0)$. That is to say, $f_y(g(y)) - f_y(x_0) = -f_y(x_0)$, or $f_y(g(y)) = 0$.

Note that in the above E and G must be isomorphic as Banach spaces. This requires, essentially, that the "number of equations" (represented by the "size" of G) be the same as the "number of unknowns" (represented by the size of E).

9. Manifolds

We have now completed our discussion of differential calculus on Banach spaces. We turn next to that mathematical object – a manifold – which be at the center of all we do hereafter. In this section, we define manifolds, give some examples, and give some constructions which yield manifolds.

A manifold is, roughly speaking, a space having the "local smoothness structure" of some Banach space. The idea, then, is to isolate, from the very rich structure of a Banach space (e.g., its vector-space structure, its norm structure) that one type of structure we call "local smoothness".

We first introduce a mechanism by which structure can be carried from Banach spaces to other things. Fix a set M and a Banach space E. A *chart* (or E–chart) on M consists of a subset U of M together with a mapping ψ from U to E, such that i) the mapping ψ is one-to-one, and ii) the subset $\psi[U]$ of E, the image of U by ψ, is open in E. Thus, a chart sets up a one-to-one correspondence between a certain subset, U, of M and a certain open subset, $\psi[U]$, of E. It is by means of this correspondence that structure is carried from E to M.

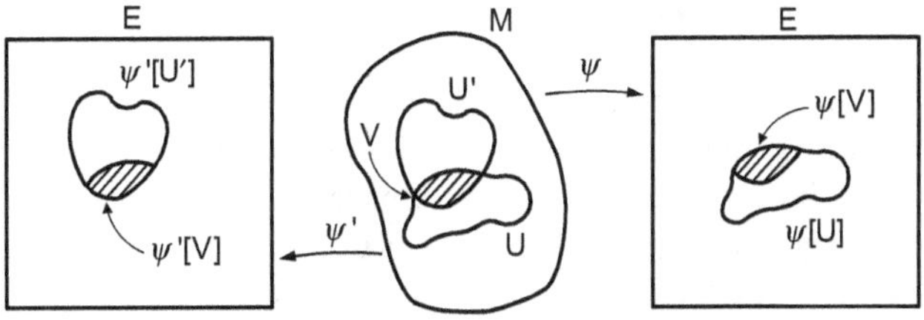

We next introduce a notion of "agreement between two charts as regards their induced smoothness structures on M". Let U, ψ and U', ψ' be two E-charts on the set M. Then on the intersection of their U's, $V = U \cap U'$, there are induced two "smoothness structures", one from ψ (which defines a correspondence between V and the subset $\psi[V]$ of E) and the other from ψ' (which defines a correspondence between V and the subset $\psi'[V]$ of E). We wish to compare these.

To this and, introduce the mapping $\psi' \cdot \psi^{-1}$ from $\psi[V]$ to $\psi'[V]$, and its inverse, $\psi \cdot \psi'^{-1}$, from $\psi'[V]$ to $\psi[V]$. These mapping represent the interaction between U, ψ, and U', ψ'. [Note that M has now been eliminated: The mappings are between subsets of Banach spaces.]

Now fix a symbol p, either a non-negative integer or the symbol "∞". We are led to the following definition: The charts U, ψ and U', ψ' on M are said to be *compatible* (or C^p–compatible) if i) $\psi[V]$ and $\psi'[V]$ are both open subsets of E, and ii) the mapping $\psi' \cdot \psi^{-1}$ from $\psi[V]$ to E, and $\psi \cdot \psi'^{-1}$, from $\psi'[V]$ to E, are both C^p mappings. The second condition is the crucial one. Note that we do not require that our mappings preserve vector-space structure, or norm structure. Rather, they need only preserve C^p differential structure. It is in this way that a single "type of structure" is isolated. Note also that two charts are necessarily compatible if their U's a fail to intersect. Example. Let U, ψ be an E–chart on M, let U' be a subset of U such that $\psi[U']$ is an open subset of $\psi[U]$, and let ψ' be ψ restricted to U'. Then U, ψ and U', ψ' are compatible.

A *manifold* consists of a non-empty set M, a Banach space E, a symbol p, and a collection ζ of E–charts on M, satisfying the following conditions:

1. Any two charts in the collection ζ are C^p compatible.

2. The charts in ζ cover M, i.e., every point of M is in at least one of the $U's$.

3. Any chart on M which is compatible with all the charts in ζ is itself in ζ.

4. The charts separate points of M in the following sense: Given distinct points p and p' of M, there are charts U, ψ and U', ψ' in ζ such that p is in U and p' in U', and such that there is a ball B centered at $\psi(p)$ in $\psi[U]$ and a ball B' centered at $\psi'(p')$ in $\psi'[U']$, with $\psi^{-1}[B]$ and $\psi'-1[B']$ not intersecting in M.

These conditions – or at least the first three – are exactly what one might have expected intuitively. The first requires that "whenever two charts in ζ induce smoothness structures in the same region of M, these structures agree". The second requires that "smoothness structure has been induced over all of M". The third guarantees that we have not induced any additional structure on M by cutting down the number of charts. Finally, the fourth condition (which is normally automatically satisfied in practice) eliminates certain pathological object, called non-Hausdorff manifolds, which are of little interest.

The manifold defined above is sometimes called a C^p manifold based on Banach space E. The charts in ζ are called the admissible charts, or just the charts. We often denote a manifold just "M", the rest understood. A subset O of manifold M is said to be *open* if, for every admissible chart U, ψ, $\psi[U \cap O]$ is open in E. These are of course, the open sets for a topology on M. The fourth condition then requires that this topological space be Hausdorff. Our third condition is often omitted, and our fourth occasionally.

One might imagine that it would be very difficult to give any examples of manifolds, for, by the third condition above, there will be an enormous collection of charts, and it might be awkward to write all these down. The possibility of giving examples easily arises from the following fact.

Lemma. Let M be a non-empty set, E a Banach space, p a symbol (a non-negative integer or "∞"), and $\hat{\zeta}$ a collection of E–charts on M satisfying conditions one, two, and four above. Denote by ζ the collection of all charts on M compatible with all those in $\hat{\zeta}$. Then ζ satisfies all four conditions.

Proof: First note that, since every chart in $\hat{\zeta}$ is necessarily in ζ, this ζ automatically satisfies the second and third conditions. Similarly, the fourth condition is also automatic. Thus, we need only verify the first. Let U, ψ and U', ψ' be two charts in ζ, and set $V = U \cap U'$. Fix point p of V, and choose chart $\hat{U}, \hat{\psi}$ in $\hat{\zeta}$, with p in $\hat{U}$ (possible, by second condition). By compatibility of $\hat{U}, \hat{\psi}$ and $U\psi$, and of $\hat{U}, \hat{\psi}$ and U', ψ', there is a ball B centered at $\hat{\psi}(p)$ such that B is in $\hat{\psi}[V]$ $(= \hat{\psi}[U]\hat{\psi}[U'])$. By compatibility of $\hat{U}, \hat{\psi}$ and U, ψ, $\psi \cdot \hat{\psi}^{-1}[B]$, a subset of $\hat{\psi}[V]$, contains a ball centered at $\psi(p)$. Thus, $\psi[V]$ is open in E, and similarly for $\psi'[V]$. Finally, since $\psi \cdot \hat{\psi}^{-1}$ and $\hat{\psi} \cdot \psi'^{-1}$ are C^p, so is their composition, $\psi \cdot \psi'^{-1}$ and similarly for $\psi' \cdot \psi^{-1}$. We conclude that U, ψ and U', ψ' are compatible.

Thus, to obtain a manifold, one need only find charts satisfying the first, second, and fourth conditions: Something which is often rather easy to do.

Examples. Let M be a C^p manifold based on E, and let $q \leq p$ (where, of course, the non-negative integers are ordered in the usual way, $q \leq \infty$ for every q, and $\infty \leq p$ only if $p = \infty$). Then, since charts C^p–compatible are also C^q–compatible, the admissible charts on M satisfy the first, second, and fourth condition with p replaced by q. By the Lemma, we obtain a C^q manifold based on E. The question of whether, given a C^q manifold and $p > q$, one can throw away some charts to get a C^p manifold is very difficult.

Example. Let E be a Banach space, and let M be the set E. Introduce a chart with $U = M$ (a subset of M), and ψ the identity mapping (from $U = M = E$ to E). This single chart satisfies the first condition, since every chart is compatible with itself, and obviously the second. This chart also satisfies the fourth condition, since given distinct points of a Banach space, one can find non-intersecting balls centered at those points. By the Lemma, we obtain a C^∞ manifold based on E.

Example. Let M denote the set of all sequences, $(r_1, r_2, \ldots)$, of real numbers with $(r_1)^2 + (r_2)^2 + \ldots = 1$. Let E be the Banach space of all sequences $(s_1, s_2, \ldots)$ the sum of the squares of whose entries converges. We introduce some E–charts on M. Let U be the subset of M consisting of all elements with $r_1 > 0$. For $(r_1, r_2, \ldots)$ in U, set $\psi(r_1, \ldots) = (r_2, r_3, \ldots)$ (an element of E, since the sum of the squares of the entries converges). We claim that this is a chart. Indeed, for $\psi(r_1, r_2, \ldots) = \psi(r'_1, r'_2, \ldots)$, we must certainly have $r_2 = r'_@$, $r_3 = r'_3$, etc. But we must also have $r_1 = r'_1$, for each is the positive square root of one minus the sum of the squares of the other entries. That is, ψ is one-to-one. We next claim that $\psi[U]$ is the subset of E consisting all $(s_1, s_2, \ldots)$ with $(s_1)^2 + \ldots < 1$, a claim which is obvious from the observation that $\psi(\sqrt{1 - \Sigma(s_i)^2}, s_1, s_2, \ldots) = (s_1, s_2, \ldots)$. But this $\psi[U]$ is an open subset of E (in fact, is the ball of radius one centered at the origin). Thus, we have a chart.

Similarly, we obtain a chart with U consisting of $(r_1, r_2, \ldots)$ with $r_1 < 0$, and the same ψ. Doing the same thing with r_2, and then with r_3, etc., we obtain still more charts on M.

We next claim that these charts satisfy conditions one, two, and four, for $p = \infty$. The second and fourth are immediate, and so we have only to check compatibility. Let us consider, e.g., the chart U, ψ (where U requires $r_1 > 0$) and U', ψ' (where U' requires $r_2 > 0$). Then $V = U \cap U'$ consists of $(r_1, r_2, \ldots)$ in M with $r_1 > 0$ and $r_2 > 0$, whence $\psi[V]$ consists of $(s_1, s_2, \ldots)$ with $(s_1)^2 + \ldots < 1$ and with $s_1 > 0$. But this is certainly an open subset of E. Similarly for $\psi'[V]$. Now, for $(s_1, s_2, \ldots)$ in $\psi'[V]$, we have $\psi^{-1}(s_1, \ldots) = (\sqrt{1 - \Sigma(s_i)^2}, s_1, s_2, \ldots)$, whence $\psi' \cdot \psi^{-1}(s_1, \ldots) = (\sqrt{1 - \Sigma(s_i)^2}, s_2, s_3, \ldots)$. We claim that this mapping $\psi' \cdot \psi^{-1}$ from $\psi[V]$ to E is C^∞. It is derivative at $(s_1, s_2, \ldots)$ in $\psi[V]$, for example, sends $(t_1, t_2, \ldots)$ in E to the element $(\theta(1 - \Sigma(s_i)^2)^{-1/2}(s_1 t_1 + s_2 t_2 + \ldots), t_2, t_3, \ldots)$ of E. [This guess is made, of course, from ordinary calculus.]

Thus, our charts satisfy the first, second, and fourth conditions. By the Lemma, we obtain a manifold. [This is the infinite-dimensional sphere. Similarly, one obtains the finite-dimensional sphere.]

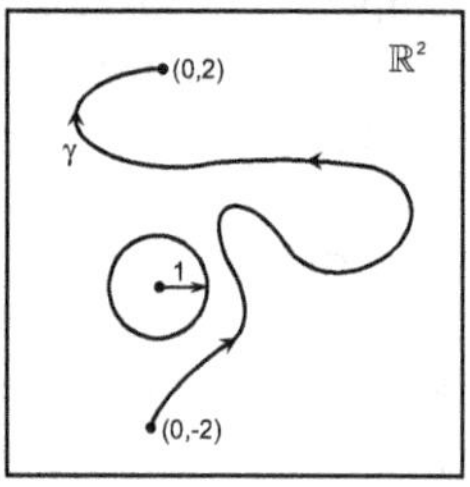

Example. Let F be the Banach space of continuous, real-valued functions f on the closed interval $[0, 1]$ with $f(0) = f(1) = 0$. Let $E = F \times F$, product of Banach spaces. Next, denote by M the set of all curves in the plane which begin at $(0, -2)$, end at $(0, +2)$, and which avoid the closed disk of radius one centered at the origin. [That is, M is the set of all continuous maps γ from $[0, 1]$ to $\mathbb{R}^2$ with $\gamma(0) = (0, -2), \gamma(1) = (0, +2)$, and with $|\gamma(r)| \leq 1$ for no r.] We introduce an E–chart on this set M. Fix an element $\gamma(0)$ of M. Let $U = M$, and let ψ be the mapping from U to E which sends γ in M to the element $\gamma(r) - \gamma_0(r)$ of E (noting that $\gamma - \gamma_0$ as a continuous map from $[0, 1]$ to $\mathbb{R}^2$ beginning and ending at the origin, is an element of E). This ψ is clearly one-to-one. Furthermore, $\psi[U]$ is open in E, for, given $\psi(\gamma)$ in E, one can find an ϵ such that, whenever $|\psi(\gamma) - \tau| \leq \epsilon$, τ is also in $\psi[U]$. Thus, this is a chart.

Again, this single chart satisfies the first, second, and fourth conditions for a manifold. By the Lemma, we obtain a manifold.

We shall later display some more interesting examples of manifolds.

There are two elementary techniques for constructing new manifolds from old. These techniques are of interest, first because they give some insight into what a manifold is, second because they yield a rich source of examples, and third because the techniques themselves often arise in practice. We now discuss these techniques.

Let M be a manifold based on Banach space E, and let N be a subset of M and F a subspace of E. We suppose that these objects are related as follows: Given any point p of N, there is an admissible chart U, ψ on M such that p is in U, and such that $\psi[U \cap N] = \psi[U] \cap F$. Of course, such a chart will not in general exist: our requirement of its existence represents an additional condition on N and F. this condition requires essentially that "N in M look like F in E".

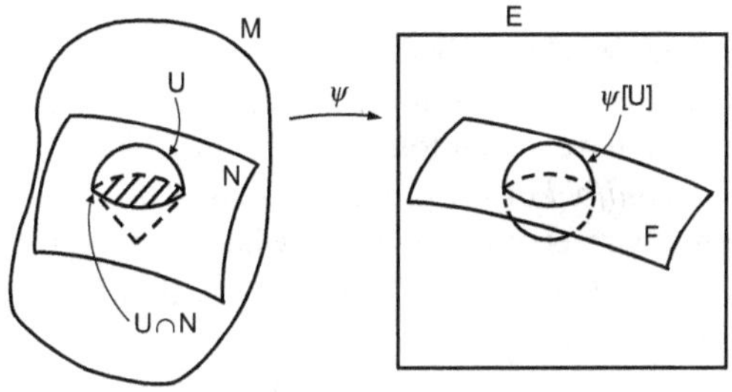

We next introduce some F–charts on this set N. Let U, ψ be an admissible chart on M, with $\psi[U \cap N] = \psi[U] \cap F$. Set $W = U \cap N$, and let φ be the restriction of ψ to W. Then W is certainly a subset of N, and φ certainly maps this W to Banach space F. Furthermore, since $\varphi[W] = \psi[U] \cap F$, this $\varphi[W]$ is an open subset of F, while, since ψ is one-to-one, so is φ. We conclude that W, φ is an F–chart on N.

The F–charts on N, so obtained, clearly satisfy the fourth condition for a manifold (since the admissible charts on M do). The second condition follows from the condition above (since the above is precisely the statement that our charts cover N). We next verify the first condition, i.e., we show that any two such charts are compatible. Let W, φ and W', φ be two F–charts on N, obtained from charts U, ψ and U', ψ' on M, respectively. Then, since $\varphi[W \cap W'] = \psi[U \cap U'] \cap F$, and since the right side is an open subset of F, so is the left side. Similarly, $\varphi'[W \cap W']$ is an open subset of F. Thus, there remains only to show that the mapping $\varphi' \cdot \varphi^{-1}$ from $\varphi[W \cap W']$ to F is C^p. To this end, fix x_0 in $\varphi[W \cap W']$, and write α for $\varphi' \cdot \varphi^{-1}$ and β for $\psi' \cdot \psi^{-1}$. Then, since β is differentiable at x_0, we have that $\beta(x) - \beta(x_0) - D\beta(x_0)(x - x_0)$ (as x ranges over points of E sufficiently close to x_0) is tangent at $x = x_0$). Now fix x in F: Then $(x - x_0)$, $\beta(x)$, and $\beta(x_0)$ are all in F, We claim that, therefore, $D\beta(x_0)(x - x_0)$ must also be in F. [Proof: Suppose not. Then there is a positive ϵ such that $|D\beta(x_0)(x - x_0) - y| \geq \epsilon$ for every y in F. Then, for all positive a, we have $|\beta(ax + (1 - a)x_0) - \beta(x_0) - D\beta(x_0)(ax - ax_0)| \geq a\epsilon$, since the first two terms are in F. But, setting, $x' = ax + (1 - a)x_0$ and letting a go to zero, this last inequality violates tangency of $\beta(x') - \beta(x_0) - D\beta(x_0)(x' - x_0)$ at $x' = x_0$.]. Denote by T the mapping $D\beta(x_0)$ restricted to

F, so T is a bounded linear mapping from F to F. Then we have, for x in F, $\beta(x) - \beta(x_0) - D\beta(x_0)(x - x_0) = \alpha(x) - \alpha(x_0) - T(x - x_0)$. Since the left side is tangent at $x = x_0$ for x ranging over E, it is also tangent for x ranging over F. Hence, the right side is also tangent at $x = x_0$, x ranging over F. Thus, α is differentiable at $x = x_0$, and $D\alpha(x_0) = T$. Similarly, if β is C^p, then α is also C^p.

Thus, we have obtained a collection of F−charts on the set N, satisfying the first, second, and forth condition for a manifold. By the Lemma, therefore, we have a manifold N based on F. The manifold so obtained is called a *submanifold* of M. [For some reason that I do not understand, this term is normally reserved for the case when F splits in E.]

Example. Let M be a a manifold based on Banach space E, and let O be an open subset of M. We claim that this O is a sub manifold of M, i.e., that O satisfies the condition on N at the top of the previous page (with $F = E$). Indeed, let p be any point of O, and let U, ψ be any admissible chart with p in U. Then, setting $U' = U \cap O$ and letting ψ' be the restriction of ψ to U', we obtain an admissible chart U', ψ'. But, for this chart, we have $\psi'[U' \cap O] = \psi'[U'] \cap E$. Thus, we obtain a manifold O based on E.

Example. Regard the manifold M of the third example on page 37 as a subset of the manifold E (consisting of $(r_1, r - 2, \ldots)$ the sum of whose squares converges). Then M is a sub manifold.

We turn, finally, to the second method for constructing manifolds from manifolds. Let M_1 and M_2 be C^p manifolds, based on Banach spaces E_1 and E_2, respectively. We obtain a new manifold. Let $M = M_1 \times M_2$, Cartesian product of sets (so an element of M is a pair, (m_1, m_2), with m_1 in M_1 and m_2 in M_2). Next, let $E = E_1 \times E_2$, product of Banach spaces. we now introduce some E−charts on this set M. Let U_1, ψ_1 and U_2, ψ_2 be admissible charts on M_1 and M_2, respectively. Set $U = U_1 \times U_2$, a subset of M (so (m_1, m_2) in M is in U provided m_1 is in U_1 and m_2 is in U_2). Let ψ be the mapping from U to E with action $\psi(m_1, m_2) = (\psi_1(m_1), \psi_2(m_2))$, noting that the object on the right is indeed in E.

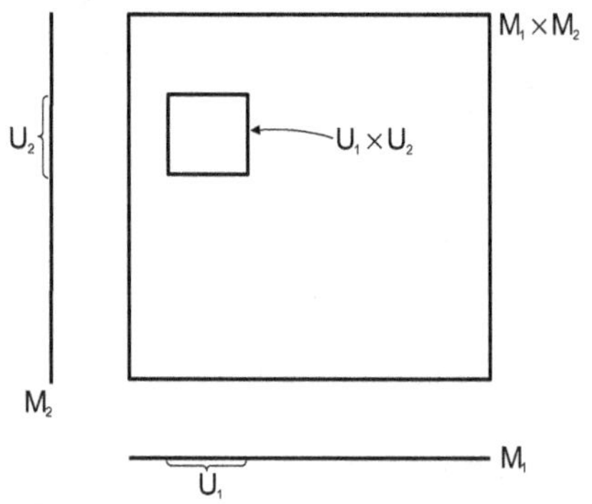

We now claim that this U, ψ is an E−chart on M. Indeed, ψ is one-to-one, for $\psi(m_1, m_2) = \psi(m'_1, m' - 2)$ implies $\psi_1(m_1) = \psi_1(m'_1)$ and $\psi_2(m_2) = \psi_2(m'_2)$, whence, since ψ_1 and ψ_2 are one-to-one, we have $m_1 = m'_1$ and $m_2 = m'_2$, whence $(m_1, m'_2) = m_2, m'_2)$. Next, note that $\psi[U]$ is open in E, for $\psi[U]$ is the subset $\psi_1[U_1] \times \psi_2[U_2]$ of E, while each factor is open in its respective Banach space. Thus, U, ψ is indeed a chart on M.

We have so far a set M, together with some charts on this set. We claim, next, that this collection of charts satisfies the first, second, and fourth conditions for a manifold. For the second condition, note that, given (m_1, m_2) in M, we have, choosing U_1, ψ_1 with m_1 in U_1 and U_2, ψ_2 with m_2 in U_2, that (m_1, m_2) is in the corresponding U. Similarly, the fourth condition follows from that condition on M_1 and M_2. Thus, we have only to verify the first condition, i.e., to show that any two of our charts are compatible. LetU, ψ and U', ψ' be two such, coming from U_1, ψ_1 and U_2, ψ_2 and U'_1, ψ'_1 and U'_2, ψ'_2, respectively. Then, setting $V_1 = U_1 \cap U'_1$ and $V_2 = U_2 \cap U'_2$, we have that $U \cap U' = V_1 \times V_2$. Thus, $\psi^{-1}[U \cap U' = \psi_1^{-1}[V_1] \times \psi_2^{-1}[V_2]$, whence, since each factor on the right is open in its Banach space, $\psi^{-1}[U \cap U']$ is open in E, and similarly for $\psi'^{-1}[U \cap U']$. Next, consider the mapping $\psi' \cdot \psi^{-1}$ from $\psi[U]$ to E. It is action on $x_1, x_2)$ in E (so x_1 is a vector in E_1 and x_2 a vector in E_2) is $\psi' \cdot \psi^{-1}(x_1, x_2) = (\psi'_1 \cdot \psi_1^{-1}(x_1), \psi'_2 \cdot \psi_2^{-1}(x_2))$. But, since $\psi'_1 \cdot \psi_1^{-1}$ (from $\psi_1[U_1]$ to E_1) is C^p and $\psi'_2 \cdot \psi_2^{-1}$ is C^p, so is this $\psi' \cdot \psi^{-1}$. We conclude that, indeed, any two of our charts are C^p–compatible.

Thus, we so far have a set M, and a collection of E–charts on M satisfying the first, second, and fourth conditions for a manifold. By the Lemma, therefore, we have a C^p manifold M based on E. This manifold is called the *product* of M_1 and M_2, written $M_1 \times M_2$.

Example, Let E_1 and E_2 be Banach spaces. Then the product of the manifold E_1 (second example, page 37) with the manifold E_2 is the manifold $E_1 \times E_2$ (product of Banach spaces).

Example Let M_1 and M_2 be C^p manifolds, and let M be the C^p manifold $M_1 \times M_2$. Fix once and for all a point $\underline{m}_2$ of M_2. Denote by N the subset of M consisting of all elements of the form $(m_1, \underline{m}_2)$ with m_1 in M_1. Then N is a submanifold of M (where the corresponding subspace F of E is just the subspace E_1 of E). Clearly, the resulting manifold N (based on E_1) is just a copy of the original manifold M_1. In this sense, then, $M_1 \times M_2$ is "sliced by submanifolds which are copies of M_1, and also by copies of M_2", just as one might expect of a product.

A characteristic feature of these two constructions should be noted. A manifold "looks locally like a certain Banach space". In each case, one simply takes a construction applicable to Banach spaces ("taking a subspace" and "taking a product", respectively) and performs essentially that same construction for manifolds, using the charts to "pull over" the construction from Banach spaces to the manifolds. This general theme persists throughout the subject; Things done to or on Banach spaces, "localized and pulled over via charts", yield things done to or on manifolds.

10. Mappings of Manifolds

Each kind of mathematical object normally comes equipped with the notion of a "structure-preserving mapping" between two such objects. For vector spaces, the mappings are linear mappings; for groups, homomorphisms, for topological spaces, continuous mappings; for Banach spaces, bounded linear mappings. Manifolds are a "kind of mathematical object". We now introduce the corresponding mappings.

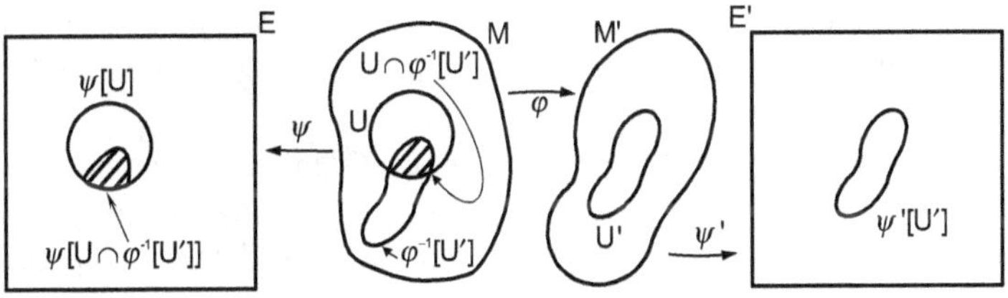

Let M and M' be C^p manifolds, based on Banach spaces E and E', respectively. Let φ be a mapping from set M to set M'. This φ is said to be a C^p *mapping* (of manifolds) provided the following condition is satisfied: Given any admissible charts U, ψ and U', ψ' on M and M', respectively, i) the subset $\psi[U \cap \varphi^{-1}[U']]$ of E is open, and ii) the mapping $\psi' \cdot \varphi \cdot \psi^{-1}$ from this $\psi[U \cap \varphi^{-1}[U']]$ to E' is C^p. We first check that these subsets and mappings make sense. First, $\varphi^{-1}[U']$ is some subset of M. Hence, $U \cap \varphi^{-1}[U']$ is some possibly smaller subset of M, and indeed is subset of U. Hence, $\psi[U \cap \varphi^{-1}[U']]$ is well-defined, and is a subset of E. For x in this subset, $\psi^{-1}(x)$ is in $U \cap \varphi^{-1}[U']$, and so, in particular, $\psi^{-1}(x)$ is in $\varphi^{-1}[U']$. Hence, $\varphi(\psi^{-1}(x))$ makes sense, and is an element of M' – indeed, is actually an element of the subset U' of M'. Hence, $\psi'(\varphi(\psi^{-1}(x)))$ makes sense and is an element of E'. Thus, everything makes sense. In condition i), let us fix U', ψ', and vary U, ψ. Then condition i) requires precisely that $\varphi^{-1}[U']$ be an open subset of M. In topological terms, a C^p mapping must be a continuous mapping of topological spaces. In the case – $p = 0$, this is the only condition, for condition ii) then follows (since compositions of continuous mappings are continuous) from condition i). For other p, however, we require still more: Roughly speaking, we require that, "if the mapping φ is pulled back via the charts to a mapping of Banach spaces (so we know what C^p means), then the result is a C^p mapping between those Banach spaces". Thus, this definition can be viewed as another instance of our general program: We know what "C^p" mapping" means for Banach spaces, and we carry over that notion to manifold via charts.

Example. Let M be a C^p manifold, and N a submanifold, so N is also a C^p manifold in its own right. Let φ be the mapping from set N to set M with $\varphi(n) = n$, i.e., which "inserts N into M". Then this φ is a C^p mapping of manifolds.

Example. Let M_1 and M_2 be C^p manifolds, so $M_1 \times M_2$ is also a C^p manifold. Let φ be the mapping from $M_1 \times M_2$ to M_1 with action $\varphi(m_1, m_2) = m_1$ ("projection onto the first factor"). Then φ is a C^p mapping of manifolds.

Example. Let M be a C^p manifold, and consider the C^p manifold $\mathbb{R}$ (a Banach space, and hence a C^∞ manifold, and hence a C^p manifold). A C^p mapping γ from R to M is called a *curve* in M.

Example Let M be a C^p manifold, and let U, ψ be an admissible chart on M. Set $K = \psi[U]$, so K is an open subset of Banach space E. We may regard K as C^p manifold (since E, as a Banach space, is a manifold, and K is an open subset of E: first example, page 40). Then ψ is a mapping from manifold K to manifold M. This is a C^p mapping.

Let M and M' be C^p manifolds, and let φ be a C^p mapping from M to M'. If φ happens to have an inverse, i.e., if there is a C^p mapping λ from M' to M such that $\lambda \cdot \varphi$ is the identity mapping on M and $\varphi \cdot \lambda$ is the identity mapping on M', then φ is called a *diffeomorphism* (or C^p diffeomorphism) from M to M'. Then necessarily λ is a diffeomorphism from M' to M. A diffeomorphism from M to M' "makes M and M' identical as manifolds" (i.e., is analogous to an isomorphism of groups or of Banach spaces, a homeomorphism of topological spaces, etc.). If there exists a diffeomorphism from M to M' then M and M' are said to be *diffeomorphic* (or C^p diffeomorphic).

Example. Consider the manifold M of the third example on page 37. Let O be the subset of M consisting of all points thereof except $(1, 0, 0, \ldots)$. Then O, as an open subset of M, is also a manifold.

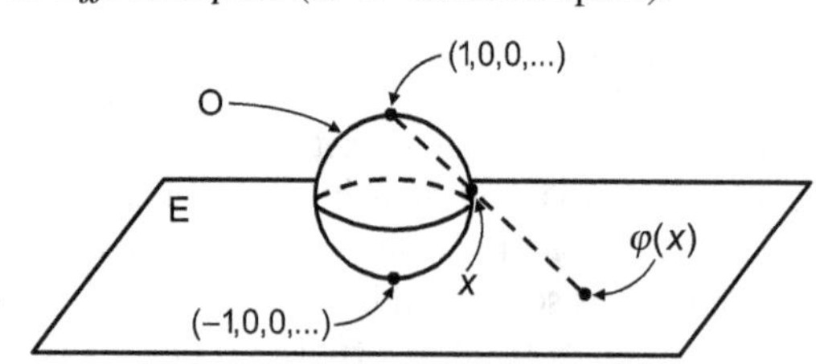

The Banach space E of that example is, as a Banach space, also a manifold. Let φ be the mapping from O to E which sends $(r_1, e_2, \ldots)$ in) to $(1/(1 - r_1))(r_2, r_3, \ldots)$ in E. This is a diffeomorphism of manifolds. [For example, a two-dimensional sphere, minus a point, is diffeomorphic with the plane.]

The composition of two C^p mappings of Banach space is C^p. As one expects, this carries over directly to manifolds.

Theorem. Let M, M', and M'' be C^p manifolds, and let φ be a C^p mapping from M to M', and φ' a C^p mapping from M' to M''. Then the mapping $\varphi' \cdot \varphi$ from M to M'' is also a C^p mapping of manifolds.

Proof: Let E, E', and E'' be the Banach spaces on which these manifolds are based. Choose charts, U, ψ, U', ψ', and U'', ψ''. Then, since $\psi'[U' \cap \varphi'^{-1}[U'']]$ is open in E', for every $U', \psi', \psi[U \cap (\varphi' \cdot \varphi)^{-1}[U'']]$ is open in E. Since the mapping of Banach spaces $\psi' \cdot \varphi \cdot \psi^{-1}$ and $\psi'' \cdot \varphi' \cdot \psi'^{-1}$ are C^p, so is their composition, $\psi'' \cdot \varphi' \cdot \varphi \cdot \psi^{-1}$.

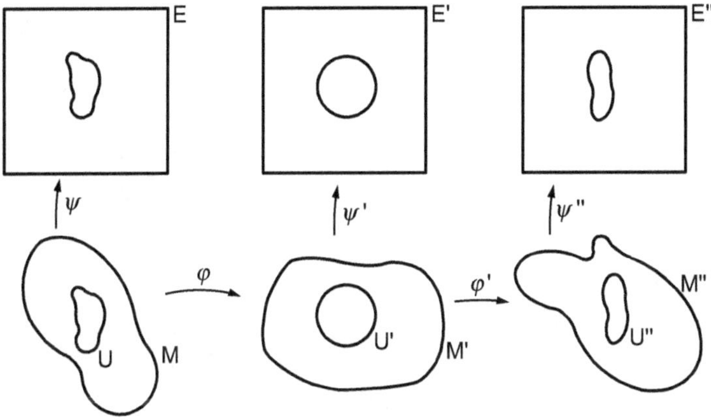

Example The composition of two diffeomorphism is a diffeomorphism, for, if λ and λ' are the inverses of φ and φ', respectively, than $\lambda \cdot \lambda'$ is the inverse of $\varphi' \cdot \varphi$, and C^p–ness is preserved under composition. Hence, "is diffeomorphic with" is an equivalence relation on manifolds.

Example. Given a curve γ on M (so γ is a C^p mapping from $\mathbb{R}$ to M), and a C^p mapping φ from M to M', then $\varphi \cdot \gamma$, as a C^p mapping from R to M', is a curve on M'.

11. Scalar Fields

We have now completed, in the last two sections, our discussion of what a manifold is and is like: We have the notion of a manifold, and of a structure-preserving mapping between manifolds. It turns out, however, that manifolds, viewed as simply objects by themselves, are not all that interesting. What is perhaps more interesting is the various objects which live naturally in the environment of a manifold. Among these objects are what will be called tensor fields. We now begin, therefore, a program of defining and finding the properties of these tensor fields. Our approach will be to first treat scalar fields and vector fields (by far the two most important examples). These two examples out of the way, we shall then look at the entire situation regarding tensor fields from a more systematic viewpoint. We begin, then, with the scalar fields.

Fix a C^p manifold M, based on Banach space E. A *scalar field* (or C^p scalar field) on M is a C^p mapping f from M to R (where R is here regarded as a C^p manifold). That is, a scalar field on M is just a real-valued function on M, a function which happens to be smooth in a certain sense.

Example. Regard Banach space E as a C^p manifold. Then a bounded linear mapping from E to $\mathbb{R}$ is a scalar field on E.

Example. Let f be a C^p scalar field on M, and γ a C^p curve. Then $f \cdot \gamma$ is one real function of one real variable (representing "evaluation of the scalar field along the curve"). As composition of C^p mappings of manifolds, this function is also C^p.

Fix C^p manifold M. Let f and f' be two C^p scalar fields on M, and consider the function $f + f''$ of M with action $(f + f'')(p) = f(p) + f'(p)$. Then this function is also a C^p scalar field (since the sum of two C^p functions on a Banach space is a C^p function). Furthermore, for a a real number, the function af with action $(af)(p) = a f(p)$ is a C^p scalar field. That is to say, the set of all C^p scalar fields on M has the structure of a real vector space. Finally, for f and f' C^p scalar fields , the function $f f'$ with action $(f f')(p) = f(p) f'(p)$ is also C^p (since true on Banach spaces). Thus, we have available a "multiplication" within our vector space. The collection of all C^p scalar fields on M thus becomes a ring (and, indeed, a commutative, associative algebra with unit).

Next, we consider the behavior of scalar fields under mappings of manifolds. Let M and M' be C^p manifolds, and let φ be a C^p mapping from M to M'. Then, given any C^p scalar field f on M', we have that $f \cdot \varphi$ is a real-valued function on M. But, as the composition of C^p mappings, this function is also C^p, and hence is a C^p scalar field on M. Thus, we can "pull back" scalar fields from M' to M under mappings from M to M'. Clearly, the "pull-back" of the sum of two fields is the sum of their pull-backs, the pull-back of a numerical multiple is that multiple of the pull-back, and the pull-back of the product is the product of the pull-backs. In other words, φ, a structure-preserving mapping from M to M', defines a structure-preserving mapping from scalar fields on M' to those on M. [In algebraic language, this last mapping is a homomorphism of rings.]

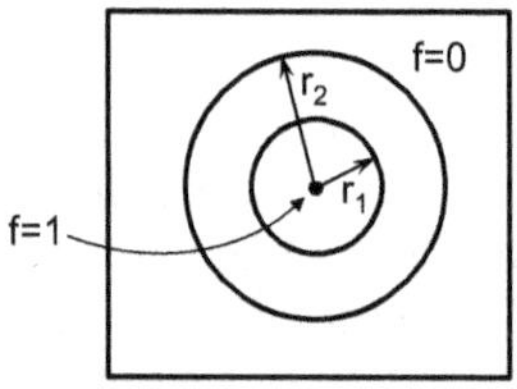

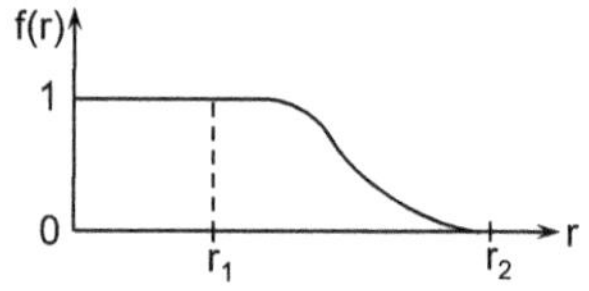

Finally, we give some examples of some interesting – and perhaps a bit surprising – difference between the properties of scalar fields in the finite – and infinite-dimensional cases. Consider first the following statement: for E a Banach space, and r_1 and r_2 positive numbers with r_1, r_2 there is a C^∞ real-valued function f on E with $f(x) = 1$ whenever $|x| \leq r_1$ and $f(x) = 0$ whenever $|x| \geq r_2$. We first note that this statement is true if E is finite-dimensional: Indeed, for $(r_1, \dots, r_n)$ in R^n, set $r = [(r_1)^2 + \dots + (r_n)^2]^{1/2}$, and let f be a C^∞ function of r which has value one for $r \leq r_1$ and zero for $r \geq r_2$. In fact, this statement in the finite-dimensional case represents an important and frequently used property of R^n: It allows one to "localize arguments" by choosing functions which do what one wants in some small region, but which, since they are zero farther away, do not do anything very nasty outside of that small region. For example, one can easily show from our fact that a finite-dimensional manifold admits a "reasonable number" of C^∞ scalar fields, in, e.g., the following sense: Given distinct points p and $p!$ of finite-dimensional M, and any numbers a and a', there exists a C^∞ scalar field f on M with $f(p) = a$ and $f(p') = a'$.

Is the statement of the previous paragraph true in general, i.e., is it also true in the infinite-dimensional case? It turns out that the answer is no.

Example. Let E be the Banach space of all sequences of reals the sum of whose absolute values converges, with norm this sum. Fix positive r, and let f be a C^1 scalar field on the ball B of radius r centered at 0. Also, let U be an open subset of E whose closure, $\overline{U}$ is in B. Then, given any point p of U and any positive ϵ, there is a point $\hat{p}$ of B, not in U, such that $|f(\hat{p}) - f(p)| \leq \epsilon$.

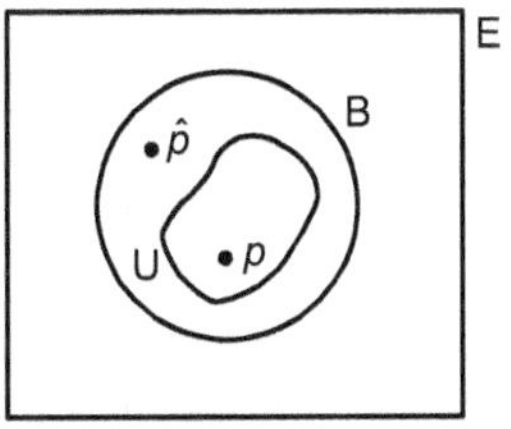

Proof: Nothing is lost be setting $r = 1$ and $f(p) = 0$. Fix, once and for all, positive number ϵ. Denote by Γ the collection of all pairs (t_0, γ), where t_0 is a non-negative number and γ is a continuous mapping from the closed interval $[0, t_0]$ to $\overline{U}$, satisfying the following conditions: i) $\gamma(0) = \beta$, and $|\gamma(t_0) - \gamma(0)| \geq t_0/2$, ii) for any t and t' in $[0, t_0]$, $|\gamma(t) - \gamma(t') \leq |t - t'|$, and iii) $f(\gamma(t_)) \leq \epsilon t_0$. [Think of the curve as a "moving point in $\overline{U}$ whose location at time t is $\gamma(t)$". The first condition then requires that "the point begin at p, and manage to achieve distance at least $t_0/2$ from p by time $t - 0$"; the second that "the speed of the moving point not exceed one"; the third that "the function f not be too large where the curve ends.] We now partially order this set Γ by the relation "is an extension of": $(t_0, \gamma) \leq (t'_0, \gamma')$ provided $t_0 \leq t'_0$ and $\gamma = \gamma'$ on $[0, t_0]$. Note that, by condition i) and the fact that $\overline{U}$ is a subset of a ball of radius one, we have $t_0 \leq 4$ for every (t_0, γ) in Γ.

We next claim that this partially ordered set Γ satisfies the condition for Zorn's Lemma. Indeed, given any totally ordered subset of Γ, we certainly obtain, by "stringing these extensions together", a t_0 (the lub of the $y'_0 s$ of the totally ordered subset), and a continuous mapping γ from $[0, t_0]$ to $\overline{U}$ satisfying the first part of condition i) and condition ii). But, by condition ii), the sequence $\gamma(t_0/2), \gamma(3t_0/4), \gamma(7t_0/8), \ldots$ in $\overline{U}$ is Cauchy, whence, since $\overline{U}$ is closed, this sequence converges to some element of $\overline{U}$. Denote this element $\gamma(t_0)$, so we now have γ a mapping from $[0, t_0]$ to $\overline{U}$. But this (t_0, γ) is in Γ, the three conditions following from continuity. Thus, we have obtained our upper bound, and so verified the condition for Zorn's Lemma.

By Zorn's lemma, there is a maximal element, (t_0, γ), of Γ. If $\gamma(t_0)$ (necessarily in $\overline{U}$ were not in U, then we would be done (for, setting $\hat{p} = \gamma(t_0)$, we would have $|f(\hat{p})| \leq \epsilon t_0 \leq 4\epsilon$, where the first inequality is from condition iii) and the second from $t_0 \leq 4$). Thus, we have only to show that assumption that $x_0 = \gamma(t_0)$ is in U leads to a contradiction. Make this assumption, and first note that, since f is C^1 at x_0, there is a positive δ such that, whenever $|z| \leq \delta, |f(x_0 + z) - f(x_0) - Df(x_0)(z)| \leq \epsilon\delta$. We next claim that there is a vector z in E such that $|z| = \delta$, $|x_0 - p + z| \geq |x_0 - p| + \delta/2$, and $Df(x_0)(z) = 0$. [Proof: Since $x_0 - p$ is in E, it is represented by a sequence, $(r_1, r_2, \ldots)$, of reals the sum of whose absolute values converges. Choose two entries of this sequence, say r_2 and r_5, with $|r_2| \leq \delta/4$ and $|r_5| \leq \delta/4$. Consider the subspace of E consisting of elements all of whose entries, except possibly the second and fifth, are zero. Since $Df(x_0)$ is a linear mapping from this subspace to the reals, and since this subspace is two-dimensional, there is some element, z, of this subspace with $|z| = \delta$ and $Df(x_0)(z) = 0$. But now, since z is of the form $(0, u, 0, 0, v, 0, \ldots)$, and since $x_0 - p = (r_1, e_2, \ldots)$ has $|r_2| \leq \delta/4$, we have also $|x_0 - p + z| \geq |x_0 - p| + \delta/2$. This z, then, is what we wanted.]

Now set $\hat{t_0} = t_0 + \delta$, and let $\hat{\gamma}$ be the mapping from $[0, \hat{t_0}]$ with $\hat{\gamma}(t) = \gamma(t)$ for t in $[0, t_0]$, and $\hat{\gamma}(t) = \gamma(t_0) + (t - t_0)/\delta$ z for $t \geq t_0$. We show that this $(\hat{t_0}, \hat{\gamma})$ is also in Γ, violating maximality of (t_0, γ), and thus giving us our desired contradiction. Condition i): Clearly, $\hat{\gamma}(0) = p$. Also, $|\hat{\gamma}(\hat{t_0}) - \hat{\gamma}(0)| = |\gamma(t_0) + z - \gamma(0)| = |x_0 - p + z| \geq |x_0 - p| + \delta/2 \geq 1/2(t_0 + \delta) = 1/2\hat{t_0}$, where we used a property of z above in the third step, and condition i) on (t_0, γ) in the fourth. Condition ii):

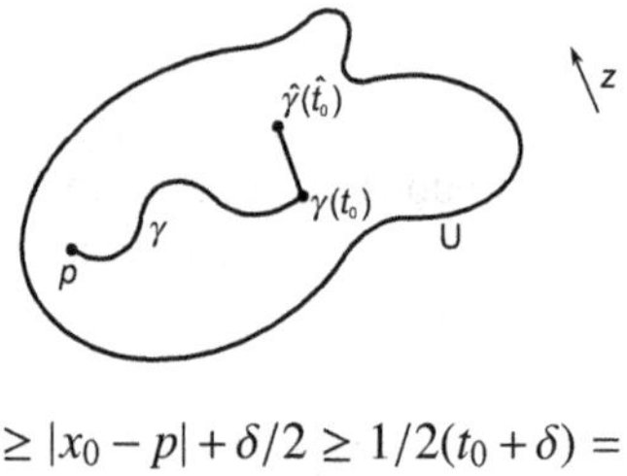

For, e.g., t in $[t_0, \hat{t_0}]$ and t' in $[0, t_0]$, we have $|\hat{\gamma}(t) - \hat{\gamma}(t')| = |\gamma(t_0) + (t - t_0)/\delta z - \gamma(t')| \leq |\gamma(t_0) - \gamma(t')| + (t - t_0)/\delta\delta \leq (t_0 - t') + (t - t_0) = |t - t'|$, where we used $|z| = \delta$ in the second step, and condition ii) on (t_0, γ) in the third. Condition iii): Since $Df(x_0)(z) = 0$ and $|z| = \delta$, we have, by the paragraph above, that $|f(x_0+z) - f(x_0)| \leq \epsilon\delta$. Hence, $|f(\hat{\gamma}(\hat{t_0}))| = |f(x_0+z)| \leq |f(x_0)| + |f(x_0 + z) - f(x_0)| \leq \epsilon t_0 + \epsilon\delta = \epsilon\hat{t_0}$. This completes the proof of the result claimed in our example.

The crucial step in the proof above is finding the "direction z, motion along which gets one significantly further from p, but along which the function f does not increase too much". The statement itself asserts that "the C^1 function f tastes outside of U any value that it assumes inside of U". This behavior of C^1 functions in infinite dimensions is rather like the well-known behavior of harmonic functions in finite dimensions. One concludes, then, that C^1 functions on certain infinite-dimensional spaces are similar in their behavior to harmonic functions on finite-dimensional spaces. Indeed, we can push this analogy still further. The only harmonic functions on a finite-dimensional sphere are the constants. In infinite dimensions, we have the following.

<ins>Example.</ins> Let E be the Banach space of the example above. Let $\tilde{E}$ be a copy of the set E (with φ a one-to-one, onto mapping from $\tilde{E}$ to E), and let M be the union of $\tilde{E}$ with one additional point ω We introduce two E–charts on the set M. For one, set $U = \tilde{E}$, and $\psi = \varphi$. For the other, set U' the subset of M consisting of ω, together with those κ in $\tilde{E}$ with $\varphi(\kappa) =\neq 0$ and let $\psi'(\omega) = 0$ and $\psi'(\kappa)\varphi(\kappa)/|\varphi(\kappa)|^2$, for κ in $\tilde{E}$. These two charts make M a C manifold based on E. [This M is "like a sphere", in, e.g., the sense that the same construction in the finite-dimensional case yields a sphere.]

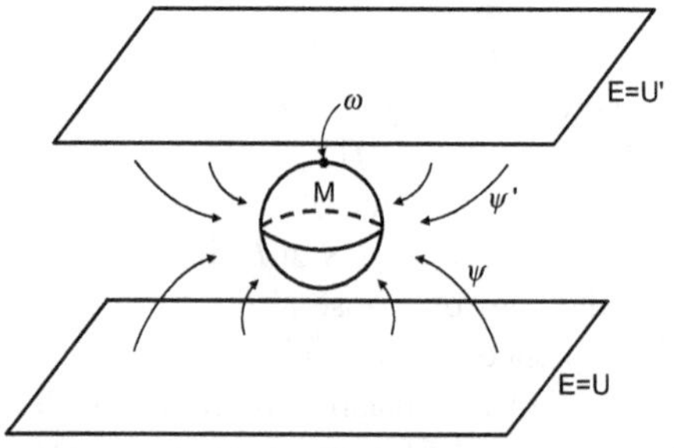

We now claim that the only C^1 scalar fields f on this manifold M are those which are constant. Indeed, let f be such a field, and let $f(\omega) = a$. Then $f \cdot \psi'^{-1}$ is a C^1 function on E, with $f \cdot \psi'^{-1}(0) = a$. Fix positive ϵ. Then, by continuity, there is a positive δ such that whenever $|x| \leq \delta$, $|f \cdot \psi'^{-1}(x) - f \cdot \psi'^{-1}(0)| \geq \epsilon$. Now, $f \cdot \psi^{-1}$ is a C^1 function on E. By what we have just shown, whenever $|x| \leq 1/\delta$, $|f \cdot \psi^{-1}(x) - a| \leq \epsilon$. But, by the result of the previous example, it follows that $|f \cdot \psi^{-1}(x) - a| \leq \epsilon$ for every x in E. Hence, $|f(p) - a| \leq \epsilon$ for every p in M. Since ϵ is arbitrary, we must have $f(p) = a$ for every p in M.

Thus, the manifold of this example, at least, has no interesting C^1 scalar field whatever.

12. Vector Fields

We have now completed our discussion of the first example of a field on a manifold: scalar fields. We turn now to the second example.

Let M be a C^p manifold ($p \geq 1$) based on Banach space E. Fix a point p of M. Consider now the collection of all pairs, $(x; U\psi)$, where x is a vector in E and U, ψ is an admissible chart on M with p in U. Given two such, write $(x; U, \psi) \approx (x'; U', \psi')$ if $D(\psi' \cdot \psi^{-1})(\psi(p)(x) = x'$. We now claim that this $\approx$ is an equivalence relation. [Proof: Clear, $(x; U, \psi)$, $\approx$ $(x; U, \psi)$. Next, let $(x; U, \psi) \approx (x'; U', \psi')$. We have that $(\psi' t \psi^{-1}) \cdot (\psi \cdot \psi'^{-1})$ is the identity mapping. Taking the derivative of this equation, using the chain rule, we have $D(\psi' \cdot \psi^{-1})(\psi(p)) D(\psi \cdot \psi'^{-1})(\psi'(p)) = I$, the identity on E. Hence, since $D(\psi' \cdot \psi^{-1})(\psi(p))(x) = x'$ we have $D(\psi \cdot \psi'^{-1})(\psi'(p))(x') = x$, i.e., $(x'; U', \psi') \approx (x; U, \psi)$. Finally, let $(x; U, \psi) \approx (x'; U', \psi')$ and $(x'; U', \psi') \approx (x''; U'', \psi'')$. Taking the derivative of $(\psi'' \cdot \psi'^{-1}) \cdot (\psi' \cdot \psi^{-1}) = \psi'' \cdot \psi^{-1}$, we have that $D(\psi'' \cdot \psi'^{-1})(\psi'(p)) D(\psi' \cdot \psi^{-1})(\psi(p)) = D(\psi'' \cdot \psi^{-1}(\psi(p))$. Applying each side to x, we have $D(\psi'' \cdot \psi^{-1})(\psi(p))(x) = D(\psi'' \cdot \psi'^{-1})(\psi'(p))(D(\psi' \cdot \psi^{-1}(\psi(p))(x)) = D(\psi'' \cdot \psi'^{-1})(\psi'(p))(x') = x''$. hence, $(x; U, \psi) \approx (x''; U'', \psi'')$.] An equivalence class is called a *tangent vector* (to M, at p), or just vector in M.

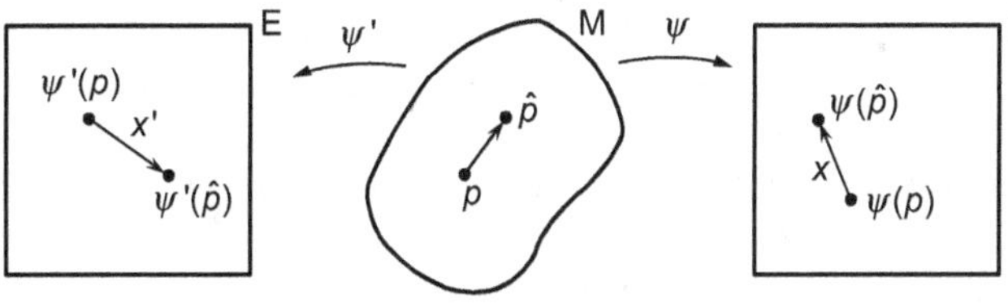

Intuitively, a tangent vector at p represents "an infinitesimal displacement from p, in M". We support this intuitive picture, and thus motivate the definition above, by the following "first order calculation". Fix point $\hat{p}$ in M "near p". Let us represent the relationship between these points in terms of chart U, ψ, with p in U. This representation can be accomplished by considering the vector $x = \psi(\hat{p}) - \psi(p)$ in E. A different chart gives a different representation, namely, U', ψ' gives $x' = \psi'(\hat{p}) - \psi'(p)$. To compare these two "representations of the same displacement in M", we note that $x' = \psi'(\hat{p}) - \psi'(p) = (\psi' \cdot \psi^{-1})(\psi(\hat{p})) - (\psi' \cdot \psi^{-1})(\psi(p)) \approx$

$D(\psi' \cdot \psi^{-1})(\psi(p))(\psi(\hat{p}) - \psi(p)) = D(\psi' \cdot \psi^{-1})(\psi(p))(x)$, where "$\eqsim$" means "ignoring a mapping tangent at $\psi(p)$". But this formula is precisely the equivalence relation above. The "term ignored becomes negligible as the displacement between $\hat{p}$ and p goes to zero". Thus, we interpret our equivalence relation as requiring that "various representation in terms of charts all correspond to the same infinitesimal displacement in M".

Further support for our intuitive picture comes from the following.

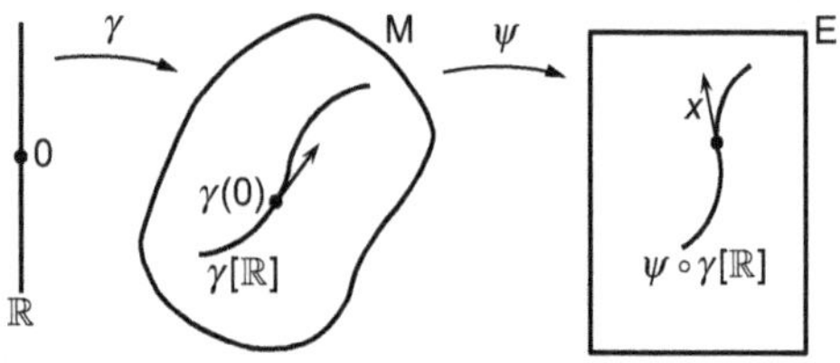

Example. Let M be a C^p manifold ($p \geq 1$), and let γ be a curve in M, i.e., a C^p mapping from $\mathbb{R}$ to M. Let p be the point $\gamma(0)$ of M. We obtain a tangent vector at p. Given any chart U, ψ in M with p in U, we have that $\psi \cdot \gamma$ is a mapping from $\mathbb{R}$ to E, whence its derivative at $0, D(\psi \cdot \gamma)(0)$ is a linear mapping from $\mathbb{R}$ to E, whence $x = D(\psi \cdot \gamma)(0)(1)$ is a vector in E. Given another such chart, we similarly set $x' = D(\psi' \cdot \gamma)(0)(1)$. But, taking the derivative of $\psi' \cdot \gamma = (\psi' \cdot \psi^{-1}) \cdot (\psi \cdot \gamma)$ at 0, we have $D(\psi' \cdot \gamma)(0) = D(\psi' \cdot \psi^{-1})(\psi \cdot \gamma(0))\, D(\psi \cdot \gamma)(0)$. applying this to the number "1" in $\mathbb{R}$, we see that $x' = D(\psi' \cdot \psi^{-1}(\psi(\gamma(0)))(x)$. That is to say, we have that $(x; U, \psi) \approx (x'; U', \psi')$. We thus obtain in this way an equivalence class as on the previous page. This tangent vector is called the *tangent* to the curve γ at $\gamma(0)$.

This last example gives what we might expect: Since a curve "moves in M", it should define, at each point of the curve, a tangent vector giving the "direction of motion in M".

Example. Let M be a C^p manifold ($p \geq 1$), p a point of M, ξ a tangent vector at p, and f a C^p scalar field on M. Let $(x; U, \psi)$ be any representative of the equivalence class ξ, and consider the number $a = D(f \cdot \psi^{-1})(\psi(p))(x)$. Given any other $(x'; U', \psi')$ in this equivalence class, we similarly define a'. But, taking the derivative of the identity $(f \cdot \psi'^{-1})(\psi' \cdot \psi^{-1}) = (f \cdot \psi^{-1})$, and using the fact that $(x; U, \psi) \approx (x'; U', \psi')$, we have that $a = a'$. Thus, we obtain, from vector ξ and scalar field f, a unique number, independent of choice of chart used to define this number. This number is called the *directional derivative* of f in the direction ξ. Intuitively, this number is " the rate of change of f under the infinitesimal displacement in M defined by ξ".

The set of all tangent vectors at p is called the *tangent space* (to M) at p. Our next goal is to determine what structure there is on this set. To this end, we first note the following fact: If ξ is any tangent vector at p, and U', ψ' is ant chart with p in U', then there is one and only one vector x' in E with $(x'; U', \psi')$ in the equivalence class ξ. Indeed, for $(x; U, \psi)$ any representative of ξ, the required unique x' is that given by $x' = D(\psi' \cdot \psi^{-1})(\psi(p))(x)$. Thus, given a chart U, ψ with p in U, we obtain a one-to-one, onto mapping α from the tangent space at p to the Banach space E (namely, $\alpha(\xi)$ is that x in E such that $(x; U, \psi)$ is in the equivalence class ξ). This correspondence α thus induces on the tangent space all the structure of the Banach space E (i.e., a vector-space structure and norm structure). We are of course interested only in that part of all this structure on the tangent space which is independent of the choice of chart. So, let U', ψ' be another. Then we obtain immediately (since, for any $\xi, (\alpha(\xi); U, \psi') \approx (\alpha'(\xi) : U'\psi')$) that $\alpha' = D(\psi' \cdot \psi^{-1})(\psi(p)) \cdot \alpha$. Clearly, then, the vector-space structure on the tangent space is chart-independent (i.e., if $\alpha(\tau) = \alpha(\xi) + \alpha(\eta)$, then $\alpha'(\tau) = \alpha'(\xi) + a\alpha'(\eta)$). The norm structure, however, does depend on the chart, for $|\alpha'(\xi)| = |D(\psi' \cdot \psi^{-1})(\psi(p))(\alpha(\xi))|$, which is not in general equal to $|\alpha(\xi)|$. We have, however, that $|\alpha'(\xi)| \leq |D(\psi' \cdot \psi^{-1})(\psi(p))| \, |\alpha(\xi)|$, and similarly, reversing the roles of α and α'. That is to say, any two norms obtained on the tangent space, from two charts, are equivalent to each other: It is only the actual numerical values of the norms which are chart-dependent. This structure on the tangent space – a vector space, together with a collection of norms theorem which make the vector space a Banach space and which are all equivalent to each other – is called a Banachable space. We can, in the tangent space, speak of sums of tangent vectors, numerical multiplies of tangent vectors, limits of a sequence of tangent vectors, Cauchy sequences of tangent vectors (all notions which depend only on the vector-space structure or on the norm up to equivalence): We cannot, for example, speak of the norm of a tangent vector (as a real number).

In the finite-dimensional case, there are a number of equivalent definitions of tangent vectors. It is of some interest to see which of these definitions agree with ours above in general (i.e., also in infinite dimensions). We consider two.

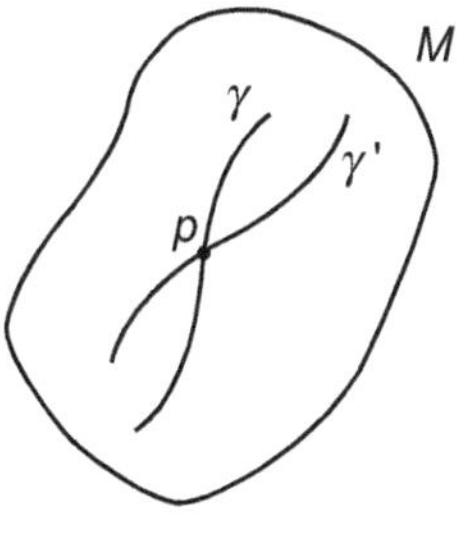

Let M be a C^p manifold ($p \geq 1$), and fix a point p of M. Given any C^p curve γ on M with $\gamma(0) = p$, and any C^p scalar field f on M, denote by $s(\gamma, f)$ the number $D(f \cdot \gamma)(0)(1)$ (i.e., the ordinary derivative, at zero, of the real function $f \cdot \gamma$ of one real variable). [In our language, this is the directional derivative of f in the direction of the tangent to γ at zero.] One then calls two curves, γ and γ', equivalent if $s(\gamma, f) = s(\gamma', f)$ for every f (i.e., if, in intuitive terms, the two curves are "tangent to each other at p"). Tangent vectors are then defined as the equivalence classes. This definition is not, in general, the same as ours, but for a minor technical reason: As we saw in the previous section, there are manifolds which admit only the constant C^p scalar fields, and for these, for example, we would obtain only one equivalence class (since we would have $s(\gamma, f) = 0$ always). We can avoid this difficulty, however, by considering instead C^p scalar fields defined on a sufficiently small open sub manifold of M containing p. [Choosing, e.g., this open subset to be the "U" of a chart, we will obtain enough C^p scalar fields.] With this one modification, the two definitions will coincide. [Any tangent vector in this sense defines one in our sense, since equivalent curves have the same tangent. Conversely, give a tangent vector in our sense, one constructs a curve whose tangent at p is that tangent vector (using a chart), and thus obtains an equivalence class of curves and hence a tangent vector in this sense.]

For the second definition, again let M be a C^p manifold ($p \geq 1$), and let p be a point of M. Denote by $\mathfrak{J}$ the collection of all C^p scalar fields on M, so $\mathfrak{J}$ is a vector space with products. A derivation on $\mathfrak{J}$ is a mapping δ from $\mathfrak{J}$ to the reals, satisfying the following conditions: i) For f and f' in $\mathfrak{J}$, $\delta(f + f') = \delta(f) + \delta(f')$, ii) For f a constant function, $\delta(f) = 0$, and iii) For f and f' in $\mathfrak{J}\delta(ff') = f(p)\delta(f') + f'(p)\delta(f)$. As an example of a derivation, we have the following: Fix a tangent vector ξ to M at p, and denote by δ the mapping from $\mathfrak{J}$ to $\mathbb{R}$ with the following action: For f in $\mathfrak{J}, \delta(f)$ is the number given by the directional derivative of f in the ξ direction. Then this δ is a derivation (the three properties above being just properties of the directional derivative). For the second definition, tangent vectors are defined as derivations on $\mathfrak{J}$. This definition also does not agree with ours, for the same technical reason as above: Not enough scalar fields in general. Again, we can avoid this difficulty by choosing for $\mathfrak{J}$ the scalar fields in some small open set containing p. This having been done, does the present coincide with our original definition? It turns out that the answer is still no.

Example. Let the manifold M be the Banach space E itself (regarded as a manifold), and let the point p be the zero vector in E. Then, since M comes equipped naturally with a chart, the tangent space at p can be identified with E. Given any scalar field f on M, denote by τ_f the element of $\mathcal{L}(E;\mathbb{R})$ which sends x in E to the real number the directional derivative of f in the x direction (regarding x as a tangent vector at 0. Then, clearly, $\tau_{f+f'} = \tau_f + \tau_{f''}$ for f a constant scalar field, $\tau_f = 0$, and $\tau_{ff'} = f(0)\,\tau_{f'} + f'(0)\tau_f$.

Now let α be any element of $\mathcal{L}(\mathcal{L}(E;\mathbb{R});\mathbb{R})$. Then for each f $\alpha(\tau_f)$ is just a real number. Hence, the mapping δ from $\mathfrak{J}$ to $\mathbb{R}$ with $\delta(f) = \alpha(\tau_f)$ is a derivative on $\mathfrak{J}$. Thus, so far we have a construction which yields, from an element α of $\mathcal{L}(\mathcal{L}(E;\mathbb{R});\mathbb{R})$, a derivation. In particular, every element x of E determines a certain α_x in $\mathcal{L}(\mathcal{L}(E;\mathbb{R}) : R)$, namely, with the following action: For τ in $\mathcal{L}(E;\mathbb{R})$ let $\alpha_x(\tau) = \tau(x)$. Thus, since an element of E yields an element of $\mathcal{L}(\mathcal{L}(E;\mathbb{R});\mathbb{R})$, and since an element of $\mathcal{L}(\mathcal{L}(E;\mathbb{R});\mathbb{R})$ yields a derivation, each element of E yields a derivation. This, of course, is just the example we remarked on in the second paragraph of this page.

Is every α in $\mathcal{L}(\mathcal{L}(E;\mathbb{R}) : \mathbb{R})$ a α_x for some x in E? If not, we shall have an example of a derivation (namely, that which comes from this α) which arises from no tangent vector. We now find, for a specific choice of E, such an α.

Let E be the Banach space of all sequences of real numbers which converge to zero, with norm the lub of the absolute values of the entries. We now claim that $\mathcal{L}(E;\mathbb{R})$ is precisely the Banach space F of all sequences of reals the sum of the absolute values of whose entries converges (with norm this sum). Indeed, for $y = (s_1, s_2, \ldots)$ in F, let τ_y be the element of $\mathcal{L}(E;\mathbb{R})$ which sends $x = (r_1, r_2, \ldots)$ in E to the number $\tau_y(x) = s_1 r_1 + s_2 r_2 + \ldots$ (noting that the sum of the right converges, since the r_i converge to zero, and the sum of the absolute values of the s_i is finite). Conversely, let τ be an element of $\mathcal{L}(E;\mathbb{R})$. Set $s_1 = \tau(1, 0, \ldots)$, $s_2 = \tau(0, 1, 0, \ldots)$, etc. Then, for any $x = (r_1, r_2, \ldots)$ in E, we must have $\tau(x) = s_1 r_1 + s_2 r_2 + \ldots$) (for the sequence in E whose n^{th} element is $(r_1, r_2, \ldots, r_n, 0, 0, \ldots)$ converges to x, whence τ of each element of this sequence must converge to $\tau(x)$. But τ of the nth element above is $s_1 r_1 + \ldots + s_n r_n$). Furthermore, this $s_1, s_2, \ldots$ must be such that the sum of their absolute values converges. [Proof: Suppose not. Then, say , we have $|s_1| + \ldots + |s_7| \geq 1, |s_8| + \ldots + |s_1 3| \geq 1$, etc. Let $r_i, i = 1, \ldots 7$ each have absolute value $1/2$, and let the sign of r_i be the same as that of the corresponding s_i; let $r_i, i = 8, \ldots 13$ each have absolute value $1/3$, each with the same sign as the corresponding s_i; etc. Then this $(r_1, r_2, \ldots)$ is in E, whereas $s_1 r_1 + s_2 r_2 + \ldots$ fails to converge: A contradiction.] Hence, $y = (s_1, s_2, \ldots)$ is an element of F. Clearly, we have $\tau = \tau_y$. We have shown, therefore, that every element of F defines an element of $\mathcal{L}(E;\mathbb{R})$, and conversely. That is, we have shown that $F = \mathcal{L}(E;\mathbb{R})$.

Now, finally, we are ready to choose our element α of $\mathcal{L}(\mathcal{L}(E;\mathbb{R}); |F|$. Let α have the following action on $\mathcal{L}(E;\mathbb{R}) = F$. For $y = (s_1, s_2, \ldots)$ in F, set $\alpha(y) = s_1 + s_2 + \ldots$ (noting that the sum converges, since the sum of the absolute values converges). For any $x = (r_1, r_2, \ldots)$ in E, the corresponding α_x has action $\alpha_x(y) = s_1 r_1 + s_2 r_2 + \ldots$. We now claim, finally, that there is no x in E such that $\alpha = \alpha_x$ (the only reasonable candidate being $x = (1, 1, 1, \ldots)$, which won't do, since this candidate is not in E). This completes our example.

The final topic, involving tangent vectors at a point, with which we shall deal is the question of their behaviour under mappings of manifolds. Let M and m' be C^p manifolds ($p \geq 1$), based on Banach spaces E and E', respectively. Let φ be a C^p mapping from M to M'. Fix a point p of M, set $p' = \varphi(p)$, and let ξ be a tangent vector to M at p. we obtain, using the mapping φ, a "corresponding" tangent vector at ξ' to M' at p'. [Intuitively, one expects such to be obtainable. Think of ξ as an "infinitesimal displacement at p in M". Then ϕ takes these "two nearby points in M to two nearby points in M', thus yielding a tangent vector at p' in M'".] The construction itself is this. Let U, ψ be a chart on M with p in U, and U', ψ' a chart on M' with p' in U'. Then $\psi' \cdot \varphi \cdot \psi^{-1}$ is a C^p mapping from an open subset of E to E'. Hence, $D(\psi' \cdot \varphi \cdot \psi^{-1})(\psi(p))$ is in $\mathcal{L}(E; E')$.

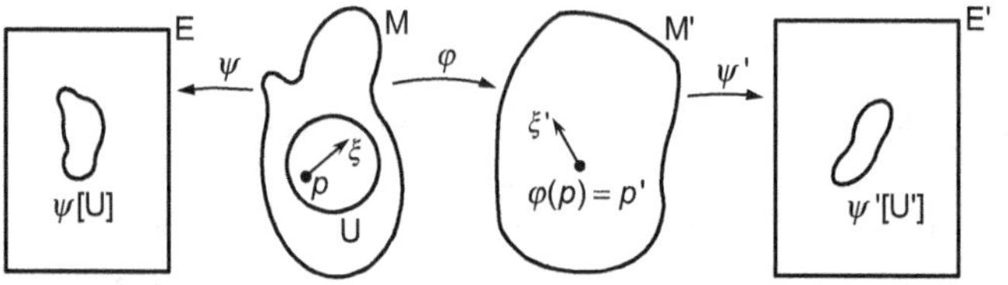

Now let x be that vector in E such that $(x; U, \psi)$ is in the equivalence class ξ. Then $x' = D(\psi' \cdot \varphi \cdot \psi^{-1})(\psi)(p))(x)$ is some vector in E'. We claim, first, that this x' is independent of the choice of the chart (U, ψ). [Indeed, given another, $\hat{U}, \hat{\psi}$, we must replace x by $\hat{x}$ such that $(\hat{x}; \hat{U}, \hat{\psi}) \approx (x; U, \psi)$. That is, we have $\hat{x} = D(\hat{\psi} \cdot \psi^{-1})(\psi(p))(x)$, which, by the chain rule, implies immediately that x' is unchanged.] Second, we claim that, if the chart U', ψ' on M' is changed to $\hat{U}', \widehat{\psi'}^{-1}$, then x' is changed to $\hat{x'} = D(\hat{\psi'} \cdot \psi'^{-1})(\psi'(p))(x')$ (again, by the chain rule and the defining equation for x'). That is to say, we have that $(\hat{x'}; \hat{U}', \hat{\psi'}) \approx (x'; U', \psi')$. Thus, we obtain an equivalence class of pairs for p' in M'. That is, we obtain a tangent vector ξ' at p'. [Note that there is only one idea, and one kind of argument, in this business, which one keeps using over and over.]

Example. Regard the reals as a manifold. Then, using the obvious chart on this manifold, we may identify a tangent vector theorem with a real number. Consider now a C^p curve γ in manifold M, with $\gamma(0) = p$. By the construction above, this C^p mapping γ of manifolds takes the tangent vector "1" at 0 in $\mathbb{R}$ to some tangent vector at p in M. This tangent vector is, of course, what we on page 54 called the tangent vector to the curve γ.

Examples. Keep in force the first two sentences of the previous example. Let f be a C^p scalar field on manifold M, and fix a point p of M. Given any tangent vector ξ at p in M, we obtain, by the construction above, since f is a C^p mapping of manifolds, a tangent vector at $f(p)$ in R. We may identify this latter whit a real number. Thus, from f and ξ we obtain a real number. This number is of course what we called the directional derivative on page 54.

It is clear from these two examples why we kept using the same argument: We were doing exactly the same thing.

The final topic to be discussed in this section is that of tangent vector fields.

Let M be a C^p manifold ($p \geq 1$). A *tangent vector field* (or vector field) on M is a mapping which assigns, to each point p of M, a tangent vector at p in M. Thus, if ξ is a tangent vector field on M, then, for each point p in M, $\xi(p)$ is an element of the tangent space at p.

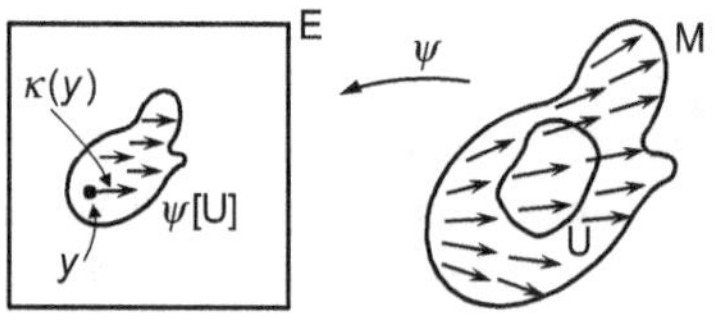

If one represents a tangent vector at p as an arrow drawn on M at p, then a tangent vector field would be a field of arrows all over M.

As with scalar fields, the fields themselves are less interesting than the smooth ones. Let ξ be any tangent vector field on C^p M, and let U, ψ be any chart. Then, for any point y of the open subset $\psi[U]$ of E, $\psi^{-1}(y)$ is some point of M,. whence $\xi(\psi^{-1}(y))$ is a tangent vector at $\psi^{-1}(y)$. Let x be that (unique) vector in E such that $(x; U, \psi)$ is in the equivalence class $\gamma(\psi^{-1}(y))$. Thus, we have constructed a mapping κ from the open subset $\psi[U]$ of E to E (y goes to $x = \kappa(y)$). Our tangent vector field ξ on M is said to be C^{p-1} if this mapping κ is C^{p-1} for every admissible chart U, ψ on M. [That is, we "pull the vector field back to E by a chart, where we know what smoothness means".]

Why do we only define C^{p-1} vector fields on C^p manifold, and not C^p fields, or even C^q fields for arbitrary q? One could certainly make such a definition, but the difficulty would be that, unless $q \leq p - 1$, the only thing satisfying that definition would be the zero vector field. The reason for this difficulty is that, for a C^p manifold, the chart-maps are only C^p related. but the equivalence relation which defines tangent vectors has in it a derivative. Hence, given a vector field on C^p M, and a chart U, ψ on M such that the "pull-back" of the field to $\psi[U]$ looks, say C^p, then, for a second chart C^p – but not C^{p+1} – related to this one, the corresponding pull-back will not look C^p. But on a C^p manifold one must admit all C^p–compatible charts (even those which are only C^p – but not C^{p+1}–related to others). Hence, given a candidate for a C^p vector field on a C^p manifold (i.e., the field looks C^p in some charts), one expects to be able to find other admissible charts in which that candidate does not look C^p. The same question does not arise for scalar fields, for there there is no "equivalence relation involving one derivative". This behaviour is perhaps not unexpected, since a tangent vector already "looks at things to first order in M", i.e., a tangent vector has already within it "one derivative". Of course, if ξ is a C^{p-1} vector field on C^p M, and if $q \leq p$, then we may also regard M as a C^q manifold, whence this same ξ will be a C^{q-1} field theorem.

We now have a set of things to study: The set of tangent vector fields on M. As usual, the study consists of finding what structure there is on this set. Let ξ and η be C^{p-1} tangent vector fields on the C^p manifold M. Then, for each point p of M, $\xi(p)$ and $\eta(p)$ are tangent vectors at p, whence their sum, $\xi(p) + \eta(p)$, is well-defined, and is also a tangent vector at p. repeating for each p, we obtain a new tangent vector field on M, which we write $\xi + \eta$. We claim that this $\xi + \eta$ is in fact C^{p-1} (immediate, since the sum of the pull-backs of ξ and η via a chart is the pull-back of the sum, on since, by the third example on page 27, the sum of C^p mappings on Banach spaces is C^p). Further, for ξ a C^{p-1} vector field on M, and a real number, the vector field with action $(a\xi)(p) = a\xi(p)$ is also C^{p-1}. Thus, the set of C^{p-1} tangent vector fields on M has the structure of a vector space.

Next, let ξ be a C^{p-1} vector field, and f a C^{p-1} scalar field, on M. Then action $(f\xi)(p) = f(p)\xi(p)$ defines another vector field on M, $f\xi$. This $f\xi$ is also C^{p-1} (since multiplication of C^{p-1} mappings of Banach spaces by C^{p-1} functions yields C^{p-1} mappings). That is, we can multiply vector fields by scalar fields. It is immediate that facts true pointwise are true for fields, i.e., $(f + f')\xi = f\xi + f'\xi$, $f(\xi + \eta) = f\xi + f\eta$. In algebraic language, the vector fields are a module over the ring $\mathfrak{F}$.

13. Tensor Products

We have now discussed two examples of fields. We shall shortly begin our program of obtaining the most general type of tensor field which can exist on a manifold. As a prerequisite for this program, we need a certain construction for obtaining Banach spaces from Banach spaces, a construction we now introduce.

Let E and $\hat{E}$ be Banach spaces. Denote by G the collection of all expressions of the form $a_1\, x_1 \otimes \hat{x}_1 + a_2\, x_2 \otimes \hat{x}_2 + \ldots$ such that i) the expression consists of a countably infinite, or possibly finite, or possibly zero, number of terms, ii) each of $a_1, a_2, \ldots$ is a nonzero real number, each of $x_1, x_2, \ldots$ a nonzero element of E, and each of $\hat{x}_1, \hat{x}_2 \ldots$ a nonzero element of $\hat{E}$, iii) for no distinct i and j does both $x_i = x_j$ and $\hat{X}_i = \hat{x}_j$ (though, of course, we could have $x_i = x_j$, provided $\hat{x}_i \neq \hat{x}_j$), iv) the sum of positive real numbers, $|a_1|\,|x_1|\,|\hat{x}_1| + |a_2|\,|x_2|\,|\hat{x}_2| + \ldots$ converges, and v) two such expressions which differ only in the order of terms are taken as representing the same element of G. [We are being a bit sloppy here. More precisely: Consider the collection of all pairs, (S, λ), where S is a set whose cardinally is at most countable, and λ is a mapping from S to the product set $\mathbb{R} \times E \times \hat{E}$, subject to appropriate conditions. Write $(S, \lambda) \approx (S', \lambda')$ is there exists a one-to-one, onto mapping ρ from S to S' such that $\lambda' \cdot \rho = \lambda$. This is an equivalence relation: The set of equivalence classes is written G.]

We next wish to introduce structure on this set G which makes it into a Banach space. We first define addition. For $a_1\, x_1 \otimes \hat{x}_1 + \ldots$ and $b_1\, y_1 \otimes \hat{y}_1 + \ldots$ in G, define their sum as follows: First, "interlace terms", i.e., write $a_1\, x_1 \otimes \hat{x}_1 + b_1\, y_1 \otimes \hat{y}_1 + a_2\, x_2 \otimes \hat{x}_2 + b_2\, y_2 \otimes \hat{y}_2 + \ldots$ if then any two terms have the same "$x \otimes \hat{x}$", i.e., if both $a_i\; x \otimes \hat{x}$ and $b_j\, x \otimes \hat{x}$ appear, replace these two terms by $(a_i + b_i)\, x \otimes \hat{x}$ if $a_i + b_j$ is nonzero, and just remove the two terms if $a_i + b_j = 0$. Doing this for all repetitions, we obtain finally an element of G. [Examples: $3\, x \otimes \hat{x} + (-3)\, x \otimes \hat{x} = 0$; $2\, x \otimes \hat{x} + 1\, (-2x) \otimes \hat{x}$ is just itself, for there are no repetitions.] Next, we define multiplication by real numbers as follows; $a(a_1\, x_1 \otimes \hat{x} + a_2\, x_2 \otimes \hat{x}_2 + \ldots) = (aa_1)\, x_1 \otimes \hat{x}_1 + (aa_2)\, x_2 \otimes \hat{x}_2 + \ldots$ (for $a \neq 0$; if $a = 0$, replace the right side by 0). This set G, with these two operations, is a vector space. [Think of the set of all expressions of the form "$x \otimes \hat{x}$", with $x \neq 0$ and $\hat{x} \neq 0$, as a "basis for G, so every vector in G can be written as a linear combination of these basis vectors". The quotes in the sentence above would be unnecessary if we had considered only finite expressions.]

Thus, we so far have constructed a vector space G. We next define a norm on this vector space as follows; $|a_1\, x_1 \otimes \hat{x_1} + \ldots| = |a_1||x_1||\hat{x_1}| + \ldots$ (noting condition iv) above). This is indeed a norm on G. [Clearly, $|a_1 x + 1 \otimes \hat{x_1} + \ldots| \geq 0$, and for equality, by ii), we must have no terms in the expression on the left. The third condition for a norm is obvious. For the second, note that in the operation of addition, one only combines or eliminates terms, and these combinations or eliminations serve only to decrease the norm of the sum of the norms.]

Thus, we now have a vector space with norm. We claim finally that this is complete, i.e., that we have a Banach. [Sketch of proof: Consider a Cauchy sequence in G, $a_1\, x_1 \otimes \hat{x_1} + \ldots, a'_1 x'_1 \otimes \hat{x'}_1 + \ldots$. Consider any expression $x \otimes \hat{x}$ which appears in at least one of the expressions representing the elements of this Cauchy sequence. Consider the sequence $r-1, r-2, \ldots$ of reals, where r_i is the coefficient of $x \otimes \hat{x}$ in the ith expression (or zero, if $x \otimes \hat{x}$ does not appear in the ith expression). Since our sequence in G is Cauchy, this sequence of real numbers must be Cauchy, whence it converges to some number r. Repeating for each choice (only a countable number need be tried) of $x \otimes \hat{x}$, we obtain a countable collection of $x \otimes \hat{x}$'s, each with a number r. Forgetting the $x \otimes \hat{x}$'s whose r's are zero, put the rest together in the expression $r\, x \otimes \hat{x} + r'\, x' \otimes \hat{x'} + \ldots$. But this expression satisfies all the conditions for membership in G. This is our candidate, then, for the element of G to which our original Cauchy sequence converges. There only remains, therefore, the (completely standard) check of convergence.]

So, we obtain a Banach space G. This G, unfortunately, is not quite the thing we are looking for. What we would really like to have be true in G is that operation $\otimes$ be distributive, and that "numerical factors can be brought inside and applied to the x's without changing the element". That is to say, we would like to have the following four equations in G be true:

$$x \otimes (\hat{x'} + \hat{x'}) - x \otimes \hat{x} - x \otimes \hat{x'} = 0$$
$$(x + x') \otimes \hat{x} - x \otimes \hat{x} - x' \otimes \hat{x} = 0$$
$$a\, x \otimes \hat{x} - (ax) \otimes \hat{x} = 0$$
$$a\, x \otimes \hat{x} - x \otimes (a\hat{x}) = 0$$

for all choices of the symbols which appear (where, of course, "$x \otimes \hat{x}$" means "$1\, x \otimes \hat{x}$", and "$-x \otimes \hat{x}$" means "$(-1)\, x \otimes \hat{x}$). But these four equations are not true in general in G: We want, therefore, to somehow "force" them to be true. This is accomplished as follows. First note that the intersection of any collection of subspaces of a Banach space is also a subspace. [Given vectors in the intersection, those vectors are in each subspace, whence any linear combination of them is in each subspace, whence that linear combination is in the intersection. So, the intersection is a vector subspace. Furthermore, the intersection, as the intersection of closed subset, is a closed subset. So, the intersection is a subspace.] Now, denote by H the intersection of all subspaces of G which contain all elements of the form of the left sides of the four equations above. This H is a subspace of G, and it consists precisely of "the things we would really rather have to zero in G". Finally, set $G/H = E \otimes \hat{E}$, a Banach space called the *tensor product* of Banach spaces E and $\hat{E}$.

We introduce the following notational conventions. Any expression satisfying the rules on page 61 defines an element of G, hence an element of G/H, hence an element of $E \otimes \hat{E}$. We shall allow ourselves to speak of such an expression as an element of $E \otimes \hat{E}$. Two elements of G whose difference is in the subspace H define the same element of G/H, and hence the same element of $E \otimes \hat{E}$. We shall represent this relationship by simply writing an equality sign between the elements (meaning "equal when regarded as elements of $E \otimes \hat{E}$"). Finally, we allow ourselves to write $0\, x \otimes \hat{x}$, a $0\, x \otimes \hat{x}$, a $0 \otimes \hat{x}$, and a $x \otimes \hat{0}$, these all being just other expression for 0 in $E \otimes \hat{E}$. With these conventions, then, the four formulae on the previous page are just true (in $E \otimes \hat{E}$). The only thing one has to be careful about, with these conventions, is the norm in $E \otimes \hat{E}$. It is not true, e.g., that the norm of $a\, x \otimes \hat{x} + a'\, x \otimes \hat{x}'$ in $E \otimes \hat{E}$ is necessarily $|a|\,|x|\,|\hat{x}| + |a'|\,|x'|\,|\hat{x}'|$ (for, e.g., the norm in $E \otimes \hat{E}$ of $x \otimes \hat{x} + (-x) \otimes \hat{x}$ is zero, since this is the zero element). Rather, the norm of such an expression is the greatest lower bound of all $|a_1|\,|x_1|\,|x\hat{x}_1| + \ldots$ with $a\, x \otimes \hat{x} + a'\, x' \otimes \hat{x}' = a_1\, x_1 \otimes \hat{x}_1 + a_2\, x_2 \otimes \hat{x}_2 + \ldots$ in $E \otimes \hat{E}$. Thus, we have in any case that, e.g., $|a\, x \otimes \hat{x} + a'\, x' \otimes \hat{x}'| \leq |a|\,|x|\,|\hat{x}| + |a'|\,|x'|\,|\hat{x}'|$. [These two remarks are just restatements of the definition of the norm in a quotient space of Banach spaces.]

The tensor product is one of those things whose definition, for some reason, is rather more complicated than the thing itself. We remark that one can, in a similar way, define the tensor product of three, or of any finite number, of Banach spaces.

We adopt the following additional convention. Let E be a Banach space, and $x_1, x_2, \ldots$ elements of E. The expression $x_1 + x_2 + \ldots$ is said to *converge* (absolutely) if the sum of real numbers, $|x_1| + |x_2| + \ldots$ converges. Then the sequence in E with n^{th} term $y_n = x_1 + \ldots + x_n$ is Cauchy (for, $n' > n$, $|y_{n'} - y_n| \leq |x_{n+1}| + \ldots + |x_{n'}|$), whence it converges to some element y of E. This y is called the *sum* of the x's, written $y = x_1 + x_2 + \ldots$. Note that these conventions are consistent with those we already have for infinite sums in tensor products.

We now give some examples of tensor products and their properties.

Example. Let E and $\hat{E}$ be finite-dimensional Banach spaces, of dimensions

n and $\hat{n}$, respectively. We "find" their tensor product. Let $x_1, \ldots, x_n$ be a basis for E, and $\hat{x}_1, \ldots, \hat{x}_{\hat{n}}$ a basis for $\hat{E}$. Then, given any $n \times \hat{n}$ matrix $a_{i\hat{i}}(i = 1, \ldots, n; \hat{i} = 1, \ldots, \hat{n})$, we can obtain an element, $\underset{i\hat{i}}{\Sigma} a_{i\hat{i}} x_i \otimes \hat{x}_{\hat{i}}$, of $E \otimes \hat{E}$. Conversely, given element $x \otimes \hat{x}$ of $E \otimes \hat{E}$, we have, expanding each of x and $\hat{x}$ in terms of the bases – $x = \Sigma r_i x_i$ and $\hat{x} = \Sigma \hat{r}_i \hat{x}_i$ – that $x \otimes \hat{x}$ is of the form above, with $a_{i\hat{i}} = r_i \hat{r}_{\hat{i}}$. Hence, every element of $E \otimes \hat{E}$ consisting of a finite sum is of this form. Hence, so is every element of $E \otimes \hat{E}$ of finite sums, and since limits of elements of the form, $\Sigma a_{i\hat{i}}\, x_i \otimes \hat{x}_{\hat{i}}$, are also of this form). Finally, note that two elements of $E \otimes \hat{E}$ written in our canonical form, $\Sigma a_{i\hat{i}}\, x_i \otimes \hat{x}_{\hat{i}}$ and $\Sigma a'_{i\hat{i}} x_i \otimes \hat{x}_{\hat{i}'}$, are equal in $E \otimes \hat{E}$ when and only when $a_{i\hat{i}} = a'_{i\hat{i}}$ for all i and $\hat{i}$. We conclude, therefore, that $E \otimes \hat{E}$ is, as a vector space, the same as the vector space of $n \times \hat{n}$ matrices. In particular, $E \otimes \hat{E}$ is $n\hat{n}$–dimensional. [Note that $E \times \hat{E}$ is $(n + \hat{n})$–dimensional.]

Example. Let E be a Banach space. We "find" the tensor product $E \otimes \mathbb{R}$. Given any element, $a_1\, x_1 \otimes b_1 + a_2\, x_2 \otimes b_2 + \ldots$ of $E \otimes \mathbb{R}$ (where the a's and b's are reals, and the x's are in E), this is equal to $(a_1, b_1)x_1 \otimes 1 + (a_2 b_2)x_2 \otimes 1 + \ldots$ and hence to $(a_1 b_1 x_1) \otimes 1 + \ldots$). But this, in turn, is equal to $(a_1 b_1 x_2 + a_2 b_2 x_2 + \ldots) \otimes 1$ (for, each n, $(a_1 b_1 x_1 + \ldots + a_n b_n x_n) \otimes 1 - (a_1 b_1 x_1) \otimes 1 - \ldots - (a_n b_n x_n) \otimes 1$ is in H, whence, since H is closed, so is the limit as n goes to infinity). Thus, every element of $E \otimes R$ can be written in the form $x \otimes 1$, for x in E and, conversely, every x in E defines an element $x \otimes 1$ of $E \otimes \mathbb{R}$.This one-to-one, onto linear mapping from $E \otimes \mathbb{R}$ to E is norm-decreasing, since $|x \otimes 1| \le |x|$, and so is an isomorphism of Banach spaces, by the open mapping theorem. [In fact, $|x \otimes 1| = |x|$.] Thus, $E \otimes \mathbb{R}$ is isomorphic with E.

Example. Let E, $\hat{E}$ and F be Banach spaces. We find an isomorphism from $\mathcal{L}(E \otimes \hat{E}; F)$ to $\mathcal{L}(E, \hat{E}; F)$. First, let α be in $\mathcal{L}(E \otimes \hat{E}; F)$. We associate, with this α element β of $\mathcal{L}(E\hat{E}; F)$, with action $\beta(x, \hat{x}) = \alpha(x \otimes \hat{x}$ (noting that this β is indeed bilinear; and that it is bounded, since α is bounded, and since $|x \otimes \hat{x}| \le |x|\,|\hat{x}|$). For the converse, let β be in $\mathcal{L}(E, \hat{E}; F)$. Then, for $a_1\, x_1 \otimes \hat{x}_1 + \ldots$ in $E \otimes \hat{E}$, set $\alpha(a_1\, x_1 \otimes \hat{x}_1 + \ldots) = a_1 \beta(x_1, \hat{x}_1) + a_2 \beta(x_2, \hat{x}_2) + \ldots$. [The sum on the right converges. We have $\Sigma |a_i \beta(x_i, \hat{x}_i)| \le \Sigma |\beta| a_i|\,|x_i|\,|\hat{x}_i| = |\beta| \Sigma |a_i|\,|x_i|\,|\hat{x}_i|$, while this last sum is finite by membership in the tensor products.] Thus, we have a one-to-one, onto linear mapping from $\mathcal{L}(E \otimes \hat{E}; F)$ to $\mathcal{L}(E, \hat{E}; F)$. Furthermore, this mapping is norm-decreasing, for, for α and β related as above and z in $E \otimes \hat{E}$, we have, by our last calculation above, that $|\alpha(z)| \le |\beta|\,|z|$, whence $|\alpha| \le |\beta|$. Hence, it is an isomorphism of Banach spaces (open mapping).

Example. Let E, F, and G be Banach spaces. We first obtain a mapping from $\mathcal{L}(E; F) \otimes G$ to $\mathcal{L}(E; F \otimes G)$. Let $\alpha_1 \otimes z_1 + \alpha_2 \otimes z_2 + \ldots$ be in the former (so the α's are in $\mathcal{L}(E; F)$, the z's in G). Associate with this the element of $(E; F \otimes G)$ which sends x in E to $\alpha_1(x) \otimes z_1 + \ldots$ (noting that this last satisfies the condition for membership in $F \otimes G$, since $\Sigma |\alpha_i(x)|\,|z_i| \le |x| \Sigma |\alpha_i|\,|z_i|$ is finite). Call this mapping (clearly linear) from $\mathcal{L}(E; F) \otimes G$ to $\mathcal{L}(E; F \otimes G)$ ψ. For $\alpha_i \otimes z_1 + \ldots$ in $\mathcal{L}(E; F) \otimes G$, and x in E, $\psi(\alpha_1 \otimes z_1 + \ldots)(x) = |\alpha_1(x) \otimes z_1 + \ldots| \leq |\alpha_1(x)|\,|z_1| + \ldots \leq |x|(|\alpha_1|\,|z_1| + \ldots)$. Hence, ψ is a bounded linear mapping, with norm less than or equal to one. It seems likely that this ψ is also one-to-one, and that its image is a closed subspace of $\mathcal{L}(E; F \otimes G)$.

We show that, however, ψ is not onto, in general. Let $F = \mathbb{R}$ and $G = E$, so ψ is a bounded linear mapping from $\mathcal{L}(E; R) \otimes E$ to $\mathcal{L}(E; E)$. If the identity in $\mathcal{L}(E; E)$ were in the domain of ψ, say were $\psi(\alpha_1 \otimes x_1 + \ldots)$, then we would have that every element of E is a (possibly infinite) linear combination of the x_i's. Thus, we have only to find a Banach space E having no countable subset linear combinations of which give every element of E. To this end, let S be any set, and let E be the Banach space of real-valued, bounded functions on S, with norm the least upper bound of the function. "The larger S, the larger E." Clearly, by choosing S to have sufficiently large cardinality, the corresponding E will have the property above. [In fact, ψ will not be onto for any infinite-dimensional E.]

For finite-dimensional E, F and G, the map of the first paragraph of this example will be an isomorphic of Banach spaces.

We remark, finally, that there is a universal definition of the tensor product, as follows. Let E and F be Banach spaces. A tensor product of E and F consists of a Banach space G, together with a bounded bilinear mapping ψ from E, F to G such that, given any Banach space G' and any bounded bilinear mapping ψ' from E, F to G', there is a unique bounded linear mapping κ from G to G' with $\psi' = \kappa \cdot \psi$. One shows, first, that, if a tensor product so defined exists, then it is unique. Then, one shows that what we have constructed is it.

14. Tensor Spaces

Our goal is to find, and learn to manipulate, the various kinds of fields which can exist on a manifold. Let, then, M be a manifold based on Banach space E. We can divide our program into two parts: i) Obtain the various Banach spaces that can be constructed from E, and unravel their structure, ii) Obtain, from each such space, a corresponding type of field on M, and carry over what structure one can from the Banach spaces to the fields. The first is essentially an algebraic problem, the second a differential one. One might suspect, therefore, that the first will be the easier (applying here the adage "algebraic is easier than differential"). This seems, however, not to be the case, and, indeed, the first problem has not, as far as I am aware, been put into a state that I would call totally satisfactory. In this section, we state the first problem; in the next, we describe some possible lines toward its solution.

Fix, once and for all, a Banach space E. Consider i) the Banach spaces E and $\mathbb{R}$, and ii) the constructions of multilinear mappings "$\mathcal{L}(,\ldots,;)$" and tensor products "$\otimes$". We consider now all Banach spaces obtained by applying the constructions ii) to the Banach spaces i), or the constructions to the Banach spaces so obtained, etc. For example, one such would be $\mathcal{L}(E \otimes (\mathbb{R} \otimes \mathcal{L}(E; R))$, $\mathcal{L}(E; E \otimes E)$; $\mathbb{R} \otimes \mathcal{L}(\mathcal{L}(\mathcal{L}(E;\mathbb{R});\mathbb{R}); E)) \otimes E$. We call these Banach spaces the *tensor spaces* (over E), and the elements of these Banach spaces *tensors* (over E). [If we wish to give more detail, an element of tensor space A will be called and A−tensor.] These definitions are inadequate for two reasons. First, the "applying constructions" business is not precise. Second, it is not made clear that, when we speak of a tensor space, we intend to refer, not only to the Banach space itself, but also to the sequence of constructions by which it arises from E. Both of these inadequacies could be resolved by proceeding as follows. We could define a tensor space as a finite sequence of symbols chosen from "E", "$\mathbb{R}$", "$\mathcal{L}$", "$\otimes$", "," ";", "(", and "), subjects to certain rules. One would then note that, if E is a Banach space, then each finite sequence gives rise to a certain Banach space. With these last five sentences as an appendix to the definition, there should be no ambiguity.

There is of course an enormous amount of structure on the tensor spaces and tensors over E. For example:

1. Natural operations on tensors. The sum of two A–tensors is well-defined (since tensor space A is a Banach space) as is the limit of a sequence of A–tensors. The tensor product of an A–tensor α and a B–tensor β can be taken, and $\alpha \otimes \beta$ is an $(A \otimes B)$–tensor. For α and A–tensor and τ a $\mathcal{L}(A; B)$–tensor, $\tau(\alpha)$ is a B–tensor. For τ a $\mathcal{L}(A; B)$–tensor and σ a $\mathcal{L}(B; C)$–tensor, $\sigma \cdot \tau$ is a $\mathcal{L}(A; C)$–tensor. With any $(A \otimes \mathbb{R})$–tensor we may associate an A–tensor (first example, page 64).

2. Preferred tensors. Some tensor spaces have proffered elements. For example, the identity in $\mathcal{L}(A; A)$ is a "preferred" $\mathcal{L}(A; A)$–tensor. So is the element of $\mathcal{L}(A, B; A \otimes B)$ which takes the tensor product; the element of $\mathcal{L}((A; B) \otimes C; \mathcal{L}(A; B \otimes C))$ of the last example on page 64; the element of $\mathcal{L}(\mathcal{L}(\mathbb{R}; A); A)$ which sends the element α of $\mathcal{L}\mathbb{R}; A)$ to the element $\alpha(1)$ of A; the element "1" of $\mathbb{R}$.

3. Naturally isomorphic tensor spaces. Some pairs of tensor spaces are just different ways of writing essentially the same thing. For example, for A a tensor space, $\mathcal{L}(\mathbb{R}; A)$ and A are naturally isomorphic; $\mathbb{R} \otimes A$ and A are naturally isomorphic. For A, B, and C tensor spaces, $\mathcal{L}(A, B; C)$ and $\mathcal{L}(A \otimes B; C)$ are naturally isomorphic.

4. Tensor spaces natural subspaces of others. For A any tensor space, A is a natural subspace of $\mathcal{L}(\mathcal{L}(A; \mathbb{R}); \mathbb{R})$. Thus, any A–tensor can also be regarded as a $\mathcal{L}(\mathcal{L}(A; \mathcal{R});)$–tensor,but not conversely. Similarly, $E \otimes \mathcal{L}(E; R)$ is a natural subspace of $\mathcal{L}(E; E)$.

5. Tensor spaces natural quotients of others. For example, for A and B tensor spaces, E may be regarded as a quotient of the tensor space $A \otimes \mathcal{L}(A; B)$ (namely, the quotient by the kernel of the natural mapping from $A \otimes \mathcal{L}(A; B)$ to B).

Problem: Organize this situation. What tensor spaces are available? Which are isomorphic to which? Which natural subspaces or natural quotients? What operations are available on tensors? Which tensor spaces have preferred elements, and what are they? One would like to cast this subject into some manageable form, in which all elementary facts look elementary, in which complicated calculations can be performed with relative ease, in which one doesn't have to continually go back and prove new things about Banach-space operations. There are of course two halves to the problem: i) make precise terms such as "natural" and "preferred", and ii) organize what is available.

15. Natural Tensors

In the previous section we posed a rather vaguely-stated problem (in which, indeed, part of the problem is to eliminate the vagueness). We now sketch a few notions which, we suggest, offer a possible line to a solution. Our basic claim is that one can make a definition which, on the one hand, gives reasonable meaning to such words as "natural" and "preferred", and, on the other, encompasses everything in which we are interested – operations, preferred elements, isomorphic tensor spaces, etc.

We first note the following facts. Let A, A', B, and B' be Banach spaces (note necessarily tensor spaces). Let α be an isomorphism from A to A', and β an isomorphism from B to B'. We define a corresponding isomorphism τ from $\mathcal{L}(A; B)$ to $\mathcal{L}(A'; B')$ as follows: For κ in $\mathcal{L}(A; B)$, $\tau(\kappa)$ is the element of $\mathcal{L}(A'; B')$ given by $\beta \cdot \kappa \cdot \alpha^{-1}$. That is, for x' in A', $\tau(\kappa)(x') = \beta \cdot \tau(\alpha^{-1}(x'))$. Similarly, given isomorphisms from A_1 to A'_1, etc, to A'_n, and from B to B', we obtain an isomorphism from $\mathcal{L}(A_1, \ldots, A_n; B)$ to $\mathcal{L}(A'_, \ldots, A'_n; B')$. [Note that we have no freedom in writing this definition.] Next, again let A and B be Banach spaces, and again let α and β be isomorphisms from A and B to Banach spaces A' and B', respectively. Then we can define an isomorphism from $A \otimes B$ to $A' \otimes B'$ as follows: For $x_1 \otimes y_1 + x_2 \otimes + \ldots$ in $A \otimes B$, $\tau(x_1 \otimes y_1 + \ldots) = \alpha(x_1) \otimes \beta(y_1) + \alpha(x_2) \otimes \beta(y_2) + \ldots$ (noting that the sum on the right converges, by boundedness of α and β, and by the fact that the sum on the left converges). Of course, these are all isomorphisms: Their inverses are obtained by the same constructions from the inverses of α and β. Thus, we can extend isomorphisms given on some Banach spaces to isomorphisms on Banach spaces constructed (by our two constructions) from these.

Now fix Banach space E. Let ι be any isomorphism from E to E, and also let ι be the identity isomorphism from $\mathbb{R}$ to $\mathbb{R}$. [This use of one letter for two things greatly conserves symbols, and causes no confusion.] By the paragraph above we obtain, given any Banach space constructed from E's and $\mathbb{R}$'s by taking multilinear mapping and tensor products, an isomorphism from this Banach space to itself. Hence, on any Banach space constructed from these by our two operations, we obtain also an isomorphism. Continuing in this way, we extend the action of ι from just E and $\mathbb{R}$ to all the tensor spaces over E.

In more detail, then, the situation is as follows. Let A and B be tensor spaces over E, and suppose that we have obtained the action of ι on A and B. Then the action of ι on $\mathcal{L}(A; B)$ is as follows: For α in $\mathcal{L}(A; B)$, $\iota(\alpha) = \iota \cdot \alpha \cdot \iota^{-1}$. Similarly, the action of ι on $A \otimes B$ is: For $x_1 \otimes y_1 + \ldots$ in $A \otimes B$, $\iota(x_1 \otimes y_1 + \ldots) = \iota(x_1) \otimes \iota(y_1) + \ldots$

To summarize, any given isomorphism on E extends in a "natural" way to isomorphisms on all the tensor space constructed from E.

The key definition is this: element α of tensor space A is called a *natural tensor* (or natural A–tensor) if $\iota(\alpha) = \alpha$ for every ι. In other words, the natural tensors remain invariant under any isomorphism induced from one on E, i.e., they in some sense "just exist, no matter what E is doing or like". [Those familiar with category theory will recognize this definition as a thinly disguised version of a natural transformation.]

Example. Each real number is a natural tensor (since ι is the identity on $\mathbb{R}$).

Example. For any tensor space A, the identity in $\mathcal{L}(A; A)$ is a natural tensor. Indeed, for τ in $\mathcal{L}(A; A)$, and for x in A, we have $\iota(\tau)$ defined by $\iota(\iota(\tau)(\iota^{-1}(x)) = \tau(x)$. But, for τ the identity, this is clearly satisfied by $\iota(\tau) = \tau$.

Example. Let A and B be any tensor spaces. then, clearly, the zero element of $\mathcal{L}(A; B)$ is a natural tensor.

Example Let A be any tensor space, and let τ be the element of $\mathcal{L}(\mathcal{L}(\mathbb{R}; A); A)$ which sends the element of α of $\mathcal{L}(\mathbb{R}; A)$ to $\alpha(1)$. Now, for any α in $\mathcal{L}(\mathbb{R}; A)$, $\iota(\alpha)$ is that element of $\mathcal{L}(\mathbb{R}; A)$ with action $\iota(\alpha)(a) = \iota(\alpha(a))$; for any τ in $\mathcal{L}(\mathcal{L}(\mathbb{R}; A); A)$, $\iota(\tau)$ has action $\iota(\tau)(\alpha) = \iota(\tau(\iota^{-1}(\alpha))$. Now let this τ be that above. Then the right side of the last equation is $\iota((\iota^{-1}(\alpha))(1))$, which, by the action of ι on $\mathcal{L}(\mathbb{R}; A)$, is $\iota(\iota^{-1}(\alpha(1))$, which equals $\alpha(1)$. Thus, $\iota(\tau)(\alpha) = \alpha(1)$. But this right side is just $\tau(\alpha)$. So, $\iota(\tau)(\alpha) = \tau(\alpha)$ for every α, i.e., $\iota(\tau) = \tau$. We have shown, therefore, that this τ is a natural tensor. Similarly, the inverse of this τ, an element of $\mathcal{L}(A; \mathcal{L}(\mathbb{R}; A))$, is a natural tensor.

Example. Let A, B and C be tensor spaces. Denote by τ the element of $\mathcal{L}(\mathcal{L}(A, B; C); \mathcal{L}(A; \mathcal{L}(B; C)))$ which sends μ in $\mathcal{L}(A, B; C)$ to the element of $\mathcal{L}(A; \mathcal{L}(B; C))$ which sends x in A to the element of $\mathcal{L}(B; C)$ which sends y in B to $\mu(x, y)$ in C. We show that this τ is a natural tensor. First note that, for μ in $\mathcal{L}(A, B; C)$, $\iota(\mu)$ has action $\iota(\mu)(x, y) = \iota(\mu(\iota^{-1}(x), \iota^{-1}(y)))$ (with x in A and y in B). Also, for ν in $\mathcal{L}(A; \mathcal{L}(B; C)), \iota(\nu)$, has action $((\iota(\nu))(x))(y) = \iota((\nu(\iota^{-1}(x)))(\iota^{-1}(y)))$. Finally, for any τ in $\mathcal{L}(\mathcal{L}(A, B; C); \mathcal{L}(A; \mathcal{L}(B; C))), \iota(\tau)$ has action $\iota(\tau)(\mu) = \iota(\tau(\iota^{-1}(\mu)))$. Applying ι^{-1} to this last equation, we have $\iota^{-1}(\iota(\tau)(\mu)) = \tau(\iota^{-1}(\mu))$. Now let τ be that tensor given in the second sentence of this example. Each side of this last equation is an element of $\mathcal{L}(A; \mathcal{L}(B; C))$: Applying each side to x in A (to obtain an element of $\mathcal{L}(B; C)$) and then to y in B (to obtain an element of C), we have $\iota^{-1}(\iota(\tau)(\mu))(x)(y) = \tau(\iota^{-1}(\mu))(x)(y)$. Using the defining equation for our particular τ on the right, this right side is $(\iota^{-1}(\mu))(x, y)$. Using now the action of ι on $\mathcal{L}(A, B; C)$, this in turn is $\iota^{-1}(\mu(\iota(x), \iota(y)))$. Using again the definition of τ, this in turn is $\iota^{-1}(\tau(\mu)(\iota(x))(\iota(y)))$. Thus, we have so far $\iota^{-1}(\iota(\tau)(\mu))(x)(y) = \iota^{-1}(\tau(\mu)(\iota(x))(\iota(y)))$. Using now on the left the action of ι on $\mathcal{L}(A; \mathcal{L}(B; C))$, this left side is $\iota^{-1}(\iota(\tau)(\mu)(\iota(x))(\iota)y)))$. Thus, we have $\iota^{-1}(\iota(\tau)(\mu)(\iota(x))(\iota(y))) = \iota^{-1}(\tau(\mu)(\iota(x))(\iota(y)))$. Since ι is an isomorphism on C, this implies $\iota(\tau)(\mu)(\iota(x))(\iota(y)) = \tau(\mu)(\iota(x))(\iota(y))$. Since x and y are arbitrary, this implies $\iota(\tau)(\mu) = \tau(\mu)$. Since μ is arbitrary, this implies $\iota(\tau) = \tau$. Thus, our τ is indeed a natural tensor. Similarly, the inverse of τ is a natural tensor. Similarly, with "A, B" replaced by more Banach spaces in $\mathcal{L}(A, B; C)$.

Example. Let A, B and C be tensor spaces. Denote by τ the element of $\mathcal{L}(\mathcal{L}(A \otimes B; C), \mathcal{L}(A, B; C))$ which sends μ in $\mathcal{L}(A \otimes B; C)$ to that element $\nu = \tau(\mu)$ of $\mathcal{L}(A, B; C)$ which sends x in A and y in B to $\nu(x, y) = \mu(x \otimes y)$. We show that this τ is a natural tensor. For μ in $\mathcal{L}(A \otimes B; C)$, $\iota(\mu)$ has action $\iota(\mu)(z) = \iota(\mu(\iota^{-1}(z)))$ (for, of course, z in $A \otimes B$). For ν in $\mathcal{L}(A, B; C)$, $\iota(\nu)$ has action $\iota(\nu)(x, y) = \iota(\nu(\iota^{-1}(x), \iota^{-1}(y)))$. For any τ in $\mathcal{L}(\mathcal{L}(A \otimes B; C); \mathcal{L}(A, B; C))$, $\iota(\tau)$ has action $\iota(\tau)(\mu) = \iota(\tau(\iota^{-1}(\mu)))$. Applying ι^{-1} to this last equation, we have $\iota^{-1}(\iota(\tau)(\mu)) = \tau(\iota^{-1}(\mu))$. Now let τ be that particular tensor given above, and apply each side of this equation to (x, y), to obtain $\iota^{-1}(\iota(\tau)(\mu))(x, y) = \tau(\iota^{-1}(\mu))(x, y)$. By definition of τ, the right side is $(\iota^{-1}(\mu))(x \otimes y)$. By the action of ι on $\mathcal{L}(A \otimes B; C)$, this in turn is $iota^{-1}(\mu(\iota(x \otimes \iota(y)))$. By the action of τ on tensor products, this in turn is $\iota^{-1}(\mu(\iota(x) \otimes \iota(y))$. By definition of τ, this is $\iota^{-1}(\tau(\mu)(\iota(x), \iota(y)))$. Thus, we have so far that $\iota^{-1}(\iota(\tau)(\mu))(x, y) = \iota^{-1}(\tau(\mu))\iota(x), \iota(y)))$. Applying the action of ι on $\mathcal{L}(A, B; C)$ to the left side, this becomes $\iota^{-1}((\iota(\tau)(\mu))(\iota(x))(\iota(y))) = \iota^{-1}(\tau(\mu)(\iota(x), \iota(y)))$. Since ι is an isomorphism, this implies $(\iota(\tau)(\mu))(\iota(x))(\iota(y)) = \tau(\mu)(\iota(x), \iota(y))$. Since x and y are arbitrary, this implies $\iota(\tau)(\mu) = \tau(\mu)$. Since μ is arbitrary, this implies $\iota(\tau) = \tau$. Thus, τ is a natural tensor. Similarly for its inverse.

These little calculations are easier they might appear. All one must do is i) make sure that, in each step, one does something new, rather than retracing the previous step, and ii) be careful not to leave out any parentheses.

We now have the notion of a natural tensor, together with some examples. We next face two issues. First, we must make a case that essentially all structure of interest on the tensor spaces can be expressed in terms of these natural tensors. Second, we must find some way easier than that of the examples above to check naturality (for the situation would be hopeless if we had to go through all that agony for each natural tensor), and we must classify them all. We discuss these two issues in turn.

For the first issue, we begin with some definitions. For $A_1, \ldots, A_n$ and B tensor spaces, a natural $\mathcal{L}(A_1, \ldots, A_n; B)$–tensor τ will be called a *natural operation.* [We regard this τ as the operation which assigns to x_1 in $A_1, \ldots, x_n$ in A_n the tensor $\tau(x_1, \ldots, x_n)$ in tensor space B.] For A and B tensor spaces, a natural $\mathcal{L}(A, B)$–tensor for which there exists a natural

$\mathcal{L}(B, A)$–tensor such that these two linear mappings are each others inverses will be called a *natural isomorphism* (from A to B). When such a natural isomorphism exists, we say that A and B are *naturally isomorphic*. We could, similarly, define a proffered tensor in tensor space A as a natural A–tensor (but we will not, since we already have the term "natural tensor" for this purpose). Finally, we could define natural subspaces and natural quotients in the obvious way (but we will not, since we will not need these terms).

Of course, the mere introduction of these terms does not a case make. What we must now do is show, by means of a large collection of examples, that the various things one would wish intuitively to call a "natural operation" "natural isomorphism", "preferred element", etc. actually arises as such from the definitions above. Further, one would like to show that each thing named above actually has the intuitive connotations of that name. We emphasize that there is nothing to prove here: It is only a matter of eliciting conviction by means of examples. We postpone, for a moment, these examples.

For the second issue, we must discover some easy way to find all the natural tensors. To this end, we first make the following two observations: i) For τ any natural $\mathcal{L}(A_1, \ldots, A_n; B)$–tensor, and $\alpha_1, \ldots, \alpha_n$ any natural A_1–, …, A_n–tensors, respectively, $\tau(\alpha_1, \ldots, \alpha_n)$ is a natural B–tensor. [Indeed, we have, by the action of ι on $\mathcal{L}(A_1, \ldots, A_n; B)$, that $\iota(\tau(\alpha_1, \ldots, \alpha_n)) = \iota(\tau)(\iota(\alpha_1), \ldots, \iota(\alpha_n))$. But, by naturality, the right side is just $\tau(\alpha_1, \ldots, \alpha_n)$.] ii) For μ any natural $\mathcal{L}(A, B)$ A–tensor and ν any natural $\mathcal{L}(B; C)$ A–tensor, $\nu \cdot \mu$ is a natural $\mathcal{L}(A; C)$–tensor (since $\iota(\nu \cdot \mu) = \iota(\nu) \cdot \iota(\mu) = \mu \cdot \mu$.). We next note that we already have, from the discussion above, the following six examples of natural tensors: 1. For any tensor space A, the zero tensor in A is natural. 2. Any real number is a natural $\mathbb{R}$–tensor. 3. For any tensor space A, the identity in $\mathcal{L}(A; A)$ is a natural $\mathcal{L}(A; A)$–tensor. 4. For any tensor space A, we have (fourth example on page 70) a natural tensor in $\mathcal{L}(\mathcal{L}(\mathbb{R}; A); A)$, and its inverse, a natural tensor in $\mathcal{L}(A; \mathcal{L}\mathbb{R}; A))$. 5. For any tensor spaces $A_1, \ldots, A_n$ and B, we have (fifth example on page 70) a natural tensor in $\mathcal{L}(\mathcal{L}(A_1, \ldots, A_n; B); \mathcal{L}(A_1, \ldots, A_{i-1}, A_{i+1}, \ldots A_n; \mathcal{L}(A_i; B))$, together with its inverse. 6. For any tensor spaces A, B, and C, we have (first example on page 71) a natural $\mathcal{L}(\mathcal{L}(A \otimes B; C); \mathcal{L}(A, B; C))$–tensor, and its inverse. Thus, we have already some examples of a few natural tensors, as well as some constructions which yield natural tensors from natural tensors. We now formulate

Conjecture. Every natural tensor is a sum of natural tensor obtained by applying the constructions i) –ii) above to the examples 1 – 6 above. If this conjecture were true, then one would have at least some control over the natural tensors, for one would have a relatively simple algorithm for obtaining them (rather then the more complicated algorithm "guess at one, and then go through all the ι's to check that guess if it is wrong, try another guess"). As its title suggests, I know of neither a proof nor a counterexample (although of course it may very well be resolved, somewhere in the literature). We can, however, at least support this conjecture by means of examples, i.e., by demonstrating that various natural tensors that come to mind can indeed be obtained by the rules laid down in the Conjecture. In my view, it would be of some interest to first resolve this conjecture, and then to begin using it (or some modification, if the conjecture should turn out to be false) to "find" all natural tensors.

Thus, we now have two issues, each of which turns to a certain extent (although, technically, in rather different ways) on examples. It is our intention, now, to support the discussion above by means of various examples. In each example, we shall find a natural tensor using the rules set forth in the conjecture (and thus support the conjecture), while at the same time many of the natural tensions we find will correspond intuitively to "natural operations", "natural isomorphisms", etc. (thus supporting the discussion of the first issue). In short, we intend each example below to serve two roles.

Example. Denote by τ the element of $\mathcal{L}(A, \mathcal{L}(A;B);B)$ with action $\tau(x,\alpha) = \alpha(x)$. We show that this τ is natural. By example 3, the identity in $\mathcal{L}(\mathcal{L}(A;B);\mathcal{L}(A;B))$ is natural. By example 5, we have a natural element of $\mathcal{L}(\mathcal{L}(\mathcal{L}(A;B);\mathcal{L}(A:B));\mathcal{L}(A,\mathcal{L}(A;B);B))$. Applying the latter to the former, we obtain τ. [Thus, "application" is a natural operation on tensor spaces.]

Example. Denote by τ the element of $\mathcal{L}(A, \mathcal{L}(A:B):B)$ with action $\tau(x,\alpha) = \alpha(x)$. We show that this τ is natural. By example 3, the identity in $\mathcal{L}(\mathcal{L}(A:B):\mathcal{L}(A:B))$ is natural. By example 5 we have a natural of $\mathcal{L}(\mathcal{L}(A:B):\mathcal{L}(A:B)):(A,\mathcal{L}(A:B):B))$. Applying the latter to the former, we obtain τ. [Thus, "application" is a natural operation on tensor spaces.]

Example. Denote by τ the element of $\mathcal{L}(\mathcal{L}(A;B),\mathcal{L}(B;C);\mathcal{L}(A;C))$ with action $\tau(\alpha,\beta) = \beta\cdot\alpha$. We show that this τ is natural. By the example above, we have a natural element of $\mathcal{L}(B;\mathcal{L}(B;C);C)$. Applying to this the natural element of $\mathcal{L}(\mathcal{L}(B,\mathcal{L}(B;C);C);\mathcal{L}(B;\mathcal{L}(\mathcal{L}(B;C);C)))$ (example 5), we obtain a natural element of $\mathcal{L}(B;\mathcal{L}(\mathcal{L}(B);C);C))$. By the example above, we also have a natural element of $\mathcal{L}(A,\mathcal{L}(A;B);B)$. Composing these two, we obtain a natural element of $\mathcal{L}(A,\mathcal{L}(A;B);\mathcal{L}(\mathcal{L}(B;C);C))$. Again applying the natural tensor of example 5 to this one, we obtain a natural element of $\mathcal{L}(A,\mathcal{L}(A;B),\mathcal{L}(\mathcal{L}(B;C);C)$. Finally, applying the natural tensor of example 5 once again, we obtain a natural element of $\mathcal{L}(\mathcal{L}(A;B),\mathcal{L}(B;C);\mathcal{L}(A;C))$. This is our τ. [Thus, "composition" is a natural operation on tensor spaces.]

Example. The tensor product of two natural tensors is natural. By example 5, we have a natural element of $\mathcal{L}(A\otimes B;A\otimes B)$. Composing with the natural tenor of element 6, we obtain a natural element τ of $\mathcal{L}(A,B;A\otimes B)$. Hence, given natural tensors in A and B, α and β, respectively, $\tau(\alpha,\beta)$ is a natural tensor (application of natural tensors to natural tensors).

Example. Denote by τ the element of $\mathcal{L}(R, A; A)$ with action $\tau(a, \alpha) = a\alpha$. We show that this τ is natural. By example 4, we have a natural element of $\mathcal{L}(A; \mathcal{L}(R; A))$. By example 5, we have a natural element of $\mathcal{L}(\mathcal{L}(A; \mathcal{L}(R; A)); \mathcal{L}(R, A; A))$. Applying the second to the first, we obtain τ, a natural element of $\mathcal{L}(R, A; A)$. [Thus, "scalar multiplication" is a natural operation on tensor spaces.]

Example. Denote by τ the element of $\mathcal{L}(A; R \otimes A)$ with action $\tau(\alpha) = 1 \otimes \alpha$. We show that this τ is natural. Applying the natural isomorphism (example 6) from $\mathcal{L}(R \otimes A; R \otimes A)$ to $\mathcal{L}(R, A; R \otimes A)$ to the identity in the former, we obtain a natural element of $\mathcal{L}(R, A; R \otimes A)$. Applying the natural isomorphism of example 5, we obtain a natural element of $\mathcal{L}(A; \mathcal{L}(R; R \otimes A))$. Composing this with the natural isomorphism (example 4) from $\mathcal{L}(R; R \otimes A)$ to $R \otimes A$, we obtain a natural element of $\mathcal{L}(A; R \otimes A)$. This is our τ.

Example Denote by τ the element of $\mathcal{L}(A; \mathcal{L}(\mathcal{L}(A; R); R))$ which sends α in A to the element of $\mathcal{L}(\mathcal{L}(A; R); R)$ which sends μ in $\mathcal{L}(A; R)$ to the number $\mu(\alpha)$. We show that this τ is natural. By the first example on page 71, we have a natural element of $\mathcal{L}(A, \mathcal{L}(A; R); R)$. Applying the natural isomorphism of example 5, we obtain our τ. [This is the operation of "inserting A into its double dual".]

Example. Denote by τ the element of $\mathcal{L}(A \otimes \mathcal{L}(A; B); B)$ with action $\tau(x_1 \otimes \mu_1 + \ldots) = \mu_1(x_1) + \ldots$. We show that this τ is natural. Applying the natural isomorphism from $\mathcal{L}(A, \mathcal{L}(A; B); B)$ to $\mathcal{L}(A \otimes \mathcal{L}(A; B); B)$ to the natural element of $\mathcal{L}(A, \mathcal{L}(A; B); B)$, we obtain our τ.

Example. Denote by τ the element of $\mathcal{L}(A \otimes B; B \otimes A)$ which "switches the order of factors". We show that this τ is natural. We have a natural isomorphism from $\mathcal{L}(A \otimes B; A \otimes B)$ to $\mathcal{L}(A, B; A \otimes B)$, and from this last to $\mathcal{L}(B, A; A \otimes B)$, and from this last to $\mathcal{L}(A; \mathcal{L}(B; A \otimes B))$ (example 5), and from this last to $\mathcal{L}(B, A; A \otimes B)$ (example 5), and from this last to $\mathcal{L}(B \otimes A; A \otimes B)$ (example 6). Composing these and applying to the identity in $\mathcal{L}(A \otimes B; A \otimes B)$, we obtain, τ. Similarly, "the tensor product is associative".

Example Denote by τ the element of $\mathcal{L}(\mathcal{L}(A; B) \otimes C; \mathcal{L}(A; B \otimes C))$ given in the last example on page 64. We show that this τ is natural. Applying example 5 to the natural element of $\mathcal{L}(B, C; B \otimes C)$, we obtain a natural element of $\mathcal{L}(B; \mathcal{L}(C; B \otimes C))$. Applying this with the natural element of $\mathcal{L}(A, \mathcal{L}(A; B); B)$, we obtain a natural element of $\mathcal{L}(A, \mathcal{L}(A; B); \mathcal{L}(C; B \otimes C))$. But this last tensor space is naturally isomorphic with $\mathcal{L}(A, \mathcal{L}(A; B), C; B \otimes C)$, which is naturally isomorphic with $\mathcal{L}(\mathcal{L}(A; B), C; \mathcal{L}(A; B \otimes C))$, which is naturally isomorphic with $\mathcal{L}(\mathcal{L}(A; B) \otimes C; \mathcal{L}(A; B \otimes C))$. Applying these isomorphisms successively to the element of the previous sentence, we obtain τ.

It is easy to think of many other, similar, example, with similar proofs.

Finally, we remark that this unsettled state of the issue of what structure there is on the tensor spaces in no way directly affects what we shall do hereafter. We shall not, for example, attempt to use in proofs things which have not been proven. The one effect that we shall see is that we will not, occasionally, be able to state our conclusion in the pretty and general way that one might wish.

16. Natural Tensors: Continued

This section is just an appendix to the previous one. We wish to discuss two examples involving tensor spaces and natural tensors. In the first, we suggest a possible answer to the question of which tensor spaces are naturally isomorphic with which others. In the second, we indicate what happens to all the tensor spaces and natural tensors in the finite-dimensional case.

Fix a Banach space E. A tensor space A over E will be said to be in *canonical form* if it satisfies the following conditions: i) No "R" appears as a factor in a tensor product, or as an entry on the left in an "$\mathcal{L}$", ii) No entry on the left in an "$\mathcal{L}$" is the tensor product, of two other tensor spaces, and iii) No entry on the right in an "$\mathcal{L}$" is of the form $\mathcal{L}(B_i, \ldots, B_n; C)$. For example, none of the tensor spaces $\mathcal{L}(E; E \otimes R)$, $\mathcal{L}(R; E \otimes E)$, $\mathcal{L}(E \otimes E; R)$, or $\mathcal{L}(E; \mathcal{L}(E; R))$ are in canonical form, while $\mathcal{L}(E, E; R)$, $\mathcal{L}(\mathcal{L}(E; E); R)$, and $\mathcal{L}(E; E \otimes E)$ are all in canonical form.

We now claim that every tensor space is naturally isomorphic to a tensor space in canonical form. The proof is quite easy. Given tensor space A, one first uses the natural isomorphism between $R \otimes B$ and B to eliminate R's in tensor products. Then, using the natural isomorphism between $\mathcal{L}(A_1, \ldots, A_n, R; B)$ and $\mathcal{L}(A_1, \ldots, A_n; B)$ one eliminates R's on the left in $\mathcal{L}$'s. Then, using the isomorphism between $\mathcal{L}(A_1, \ldots, A_n, B \otimes C; D)$ and $\mathcal{L}(A_1, \ldots, A_n, B, C; D)$, one eliminates tensor products on the left in $\mathcal{L}$'s. Finally, using example 5, one eliminates $\mathcal{L}$'s on the right in $\mathcal{L}$'s. In this way, one obtains a sequence of natural isomorphisms which carry one from A to a tensor space in canonical form. Their composition is therefore the desired natural isomorphism.

This notion of a canonical form would not be very useful if we did not wish to claim some sort of converse to the result just proved. Specifically, we have:

Conjecture. Two tensor spaces in canonical form are naturally isomorphic if and only they differ only in i) order of entries on the left in $\mathcal{L}$'s, and ii) order of, and parenthesis about, factors in tensor products.

The "if" part is of course obvious.

It would be nice to have a proof of this conjecture. Then, not only would every tensor space be naturally isomorphic to one in canonical form, but, furthermore, a given tensor space would be naturally isomorphic to only a single tensor space in canonical form (up to the ambiguity in the conjecture) (for, were A naturally isomorphic to two in canonical form, these two would be naturally isomorphic to each other). Thus, one would have a unique representative of each tensor space. One could then, for example, worry only about natural tensors in tensor spaces in canonical form. Although a proof of this conjecture from scratch appears not to be easy, there is, I should think, a reasonable chance that one could demonstrate that it is a corollary of the conjecture in the previous section (since that earlier conjecture claims to give all the natural tensors.).

In a similar way, one could try to formulate conjectures about what natural tensors there are, what natural subspaces, etc.

We consider, finally, the special case of all this when the Banach space E is finite-dimensional. In this case, everything is known, and everything is relatively easy: It is just what one learns in linear algebra. We shall here attempt only to indicate how the tensor spaces reduce to a simple form. We begin with the following observations: If A and B are finite-dimensional, then so is $A \otimes B$; If $A_1, \ldots, A_n$ and B are finite-dimensional, then so is $\mathcal{L}(A_1, \ldots, A_n; B)$. Thus, all the tensor spaces, when E is finite-dimensional, are finite-dimensional.

We first associate, with each tensor space, a pair of non-negative integers, according to the following rules. With E, associate $(1, 0)$; with R, associate $(0, 0)$; if with A there is associated (p, p'), and with B (q, q'), associate with $A \otimes B$ $(p + q, p' + q')$; if with $A_1, \ldots, A_n$ there are associated $(p_1, {p'}_1), \ldots, (p_n, {p'}_n)$, and with B there is associated (q, q'), then associate with $\mathcal{L}(A_1, \ldots, A_n; B)$ $(q + {p'}_1 + \ldots + {p'}_n, q' + p_1 + \ldots, +p_n)$, Using these rules, proceeding inductively, we associate a pair of integers with each tensor space. The pair of integers associated with tensor space A is called the *rank* of A. For example, the rank of the tensor space given in the middle of page 67 is $(4, 4)$.

The basic statement in the finite-dimensional case is this: Two tensor spaces over finite-dimensional E are naturally isomorphic if and only if they have the same rank. We shall actually prove slightly more than this. First, denote by E' the tensor space $\mathcal{L}(E; R)$. This E' is called the *dual* of Banach space E, and has rank $(0, 1)$. Given any pair (p, q) of non-negative integers, the tensor space $\mathcal{L}(E, \ldots, E, E', \ldots, E'; R)$, with pE''s and qE's, has rank (pq). Thus, in this way we obtain a particular tensor space of each rank. We shall show that, in the finite-dimensional case, any tensor space A of rank (p, q) is naturally isomorphic to the one just displayed.

We first claim the following: For A, B, and C any finite-dimensional tensor spaces, there is a natural isomorphism from $\mathcal{L}(A; B \otimes C)$ to $\mathcal{L}(\mathcal{L}(B; A); C)$. Proof: Choose bases for A, B, and C, say $x_1, \ldots, x_m, y_1, \ldots y_n$, and $z_1, \ldots, z_s$, respectively. Then an element of $B \otimes C$ is defined by an n s matrix a_{jk} ($j = 1, \ldots, n; k = 1, \ldots, s$). Hence, an element of $\mathcal{L}(A; B \otimes C)$ is defined by a matrix a_{ijk} ($i = 1, \ldots, m; j = 1, \ldots, n; k = 1, \ldots, s$), namely, such a matrix gives the element of $\mathcal{L}(A; B \otimes C)$ which sends $x = \Sigma_i b_i x_i$ in A to $\Sigma_i b_i a_{ijk}$ in $B \otimes C$. Similarly, an element of $\mathcal{L}(B; A)$ is represented by a matrix d_{ij} (i.e., sending $y = \Sigma_j b_j y_j$ in B to $\Sigma_{ij} b_j d_{ij} x_i$ in A). Hence, an element of $\mathcal{L}(\mathcal{L}(B; A); C)$ is represented by a matrics d_{ijk}. Thus, each of the tensor spaces $\mathcal{L}(A; B \otimes C)$ and $\mathcal{L}(\mathcal{L}(B; A); C)$ is represented by matrices, where these matrices are of the same kind. We thus obtain an isomorphism between these tensor spaces by comparing the matrix representations of their elements with respect to our bases for A, B and C. One easily checks that this isomorphism is independent of bases, and that it is natural. [It is interesting to note that, in the infinite-dimensional case, neither of $\mathcal{L}(A; B \otimes C)$ and $\mathcal{L}(\mathcal{L}(B; A); C)$ is even a subspaces of the other.]

Thus, in the finite-dimensional case, we have not only all the natural isomorphisms of the previous section, but also this additional one: $\mathcal{L}(A; B \otimes C)$ is naturally isomorphic with $\mathcal{L}(\mathcal{L}(B; A); C)$. From this, we obtain the following additional isomorphisms.

1. The tensor space $B \otimes C$ is naturally isomorphic with $\mathcal{L}(B'; C)$ (where prime denotes dual). Proof: Setting $A = R$ in the isomorphism above, we have that $\mathcal{L}(R; B \otimes C)$ is naturally isomorphic with $\mathcal{L}(\mathcal{L}(B; R); C)$. But $\mathcal{L}(R; B \otimes C)$ is naturally isomorphic with $B \otimes C$, while $\mathcal{L}(B; R)$ is just B'.

2. The tensor space B is naturally isomorphic with B'' (the dual of its dual). Setting $A = R$ and $C = R$ in the isomorphism above, we have that $\mathcal{L}(R; B \otimes R)$ is naturally isomorphic with $\mathcal{L}(\mathcal{L}(B; R); R)$. But the former is naturally isomorphic with B, and the latter is B''.

3. The tensor space $\mathcal{L}(A; B)$ is naturally isomorphic with $\mathcal{L}(\mathcal{L}(B; A); R)$. Proof: Set $C = R$ in the isomorphism above.

4. The tensor space $\mathcal{L}(A; B)$ is naturally isomorphic with $\mathcal{L}(A, B'; R)$. Proof: Substituting the result of (1) above in our basic isomorphism, we have that $\mathcal{L}(A; \mathcal{L}(B'; C))$ is naturally isomorphic with $\mathcal{L}(\mathcal{L}(B; A); C)$. But the first is naturally isomorphic with $\mathcal{L}(A, B'; C)$. Now set $C = R$. Then we have that $\mathcal{L}(\mathcal{L}(B; A); R)$ is naturally isomorphic with $\mathcal{L}(A, B'; R)$. By result (3) above, therefore, we have that $\mathcal{L}(A; B)$ is naturally isomorphic with $\mathcal{L}(A, B'; R)$.

Our claim, that every tensor space is the finite-dimensional case is naturally isomorphic to one of the form given at the bottom of the previous page, is now immediate. Given a tensor space, we first eliminate all tensor products by result (1) above. All $\mathcal{L}$'s which appear on the left in $\mathcal{L}$'s are than eliminated by (3). Next, anything except "R" appearing on the right in an "$\mathcal{L}$" is eliminated by (4). Finally, multiple duals are eliminated by (2). After all these eliminations, the only possible form is that given at the bottom of page 78. Furthermore, all of these "eliminations" preserve rank, as one check directly from (1) – (4). We conclude, therefore, that every tensor space, in finite dimensions, is naturally isomorphic to the one of the form $\mathcal{L}(E, \dots, E, E', \dots, E'; R)$ of the same rank.

One could now continue in this way, finding all the natural tensors in the finite-dimensional case, etc. However, since there are no new ideas, since this amounts essentially only to restating what one knows from linear algebra in a slightly different terminology, and since we are not here particularly interested in finite dimensions, we go no further.

17. Tensor Fields

The examples (Sections 11 and 12) and the algebra (Sections 13, 14, 15, and 16) out of the way, we now turn to tensor fields on manifolds.

Let M be C^p ($p \geq 1$) manifold based on Banach space E. Fix a point p of M, and a tensor space A over E. Consider now pairs $(\alpha; U, \psi)$, where α is an element of the tensor space A, and U, ψ is an admissible chart on M, with p in U. Given two such, we write $(\alpha; U, \psi) \approx (\alpha'; U', \psi')$ if $\alpha' = \iota(\alpha)$, where ι is the isomorphism on tensor space A which arises (Section 15) from the isomorphism $\iota = D(\psi' \cdot \psi^{-1})(\psi(p))$ on E. This "$\approx$ " is an equivalence relation. [The proof consists of a word-for-word repetition of the proof for tangent vectors on page 53, together with the observation that compositions and inverses of isomorphisms on E yield, when extended to the tensor space A, the compositions and inverses of the corresponding extension.] An equivalence class will be called an *A–tensor at p*. [Note that an A–tensor is just an element of the tensor space A, i.e., an element of a certain Banach space constructed from E using multilinear mappings and tensor products, while an A–tensor at p is an equivalence class of pairs, the first entry of each of which is an A–tensor. Sometimes, when we wish to emphasize the distinction, we shall call an element of the tensor space A a *free A–tensor*.]

We next note the following fact. Given any A–tensor ξ at p, and any admissible chart U, ψ with p in U, there is one and only one free A–tensor α with $(\alpha; U, \psi)$ in the equivalence class ξ. Indeed, letting $(\alpha; U', \psi')$ be any representative of the equivalence class ξ, the desired unique α is given by $\alpha = \iota^{-1}(\alpha')$, where the isomorphism ι on A comes from $D(\psi' \cdot \psi^{-1})(\psi(p))$ on E. This free A–tensor α will be called the *component* of ξ with respect to the chart U, ψ. [Motivation for the terminology: In the finite-dimensional case, one normally chooses a basis for the tensor space A, and thereby expressed the free A–tensor α in terms of n real numbers. Thus, an A–tensor at p defines, once a chart is given, n real numbers, numbers which are normally called the components of ξ. We do not choose bases, and hence replace these components by a single α in A.]

Example. Let τ be any natural tensor in tensor space A. Consider the collection of all pairs, $(\tau; U, \psi)$ whose first entry is this natural tensor. Since τ

is ι–invariant, any two such pairs are equivalent. Thus, we obtain an equivalence class, i.e., an A–tensor at p. This A–tensor at p has, of course, the property that its component with respect to any chart is just the free A–tensor τ. A–tensors at p so obtained will be called *natural A–tensors at p.*

What structure is there on the set of A–tensors at p? Picking any chart U, ψ with p in U, we obtain a one-to-one correspondence between the set of A–tensors at p and the tensor space A. By means of this correspondence, the entire structure of A – i.e., its structure as a Banach space – can be carried over to the set of A–tensors at p. We are interested only in that part of the structure which is chart-independent. Thus, exactly as with tangent vectors, the set of A–tensors at p has the structure of a Banachable space (real vector space, many equivalent norms all of which make it a Banach space). It turns out that, in fact, there is still more structure on the tensors at p. let τ be any natural operation, e.g., a natural element of the tensor space $\mathcal{L}(A_1 \ldots, A_n : B)$. Denote by σ the corresponding (example above) $\mathcal{L}(A_1, \ldots, A_n; B)$–tensor at p. Next, let $\kappa_1, \ldots, \kappa_n$ be A_1–$, \ldots, A_n$–tensors at p, respectively, Choose a chart, and let $\alpha_1, \ldots, \alpha_n$ be their respective components. Then $\tau(\alpha_1, \ldots, \alpha_n)$ is a free B–tensor. Now change the chart $(U, \psi$ to U', ψ'. Then the components change to $\alpha_1 = \iota^{-1}(\alpha'_1), \ldots, \alpha_n = \iota^{-1}(\alpha'_n)$, while the components of σ does not change. We have $\iota^{-1}(\tau(\alpha_1, \ldots, \alpha_n)) = (\iota^{-1}(\tau))(\iota^{-1}(\alpha_1), \ldots, \iota^{-1}(\alpha_n)) = \tau(\alpha'_1, \ldots, \alpha'_n)$, where the first step is the action of ι on $\mathcal{L}(A_1, \ldots, A_n; B)$, and the second is definitions. But this equation is precisely the statement that $(\tau(\alpha_1, \ldots, \alpha_n) : U, \psi)$ is equivalent to $\tau(\alpha'_1, \ldots, \alpha'_n); U', \psi')$. Thus, we obtain a B–tensor at p. In short, natural operations are extended in the obvious way from operations on free tensors to operations on tensors at p. We shall continue to use the term *natural operation* for these operations on tensors at p.

Example. "Apply", "compose", and "take the tensor product" are natural operations on tensors at p.

Our M continues to be a C^p $(p \geq 1)$ manifold based on Banach space E. We next define fields. Let A be a tensor space over E. An A–field (on M) is a mapping which assigns, to each point p of M, an A–tensor at p. As usual, it is the smooth ones which are of most interest. Let ξ be an A–field on M, and let U, ψ be any admissible chart on M.

Then, for each point x of $\psi[U], \psi^{-1}(x)$ is a point of M (in fact, of U), whence $\xi(\psi^{-1}(x))$ is an A–tensor at $\psi^{-1}(x)$, whence its components with respect to the chart U, ψ is an element of the Banach space A. Thus, we obtain a mapping from the open subset $\psi[U]$ of E to the Banach space A. The A–field ξ is said to be C^{p-1} if this mapping is C^{p-1} for every admissible chart U, ψ.

Example. Since ι is the identity on $\mathbb{R}$, we have that, for a, and a' reals, $(a : U\psi)$ $(a'; U', \psi')$ if and only if $a = a'$. Thus, $\mathbb{R}$–tensors at p can be identified with real numbers. An $\mathbb{R}$–field on M is therefore a real-valued function on M. A C^{p-1} R–field on M is thus what we called a C^{p-1} scalar field in Section 11.

Example. A C^{p-1} E–field on M is what we called a C^{p-1} tangent vector field in Section 12.

Example. Let τ be any natural element of the tensor space A. Denote by σ the A–field on M such that, for each p in M, $\iota(p)$ is the A–tensor at p associated with this τ. This A–field is C^{p-1} (for the mapping above from $\psi[U]$ to A sends all of $\psi[U]$ to the single element τ of A, and this constant mapping is certainly C^{p-1}).

The (pointwise) sum of two C^{p-1} A–fields is a C^{p-1} A–field (since sums of C^{p-1} mappings of Banach spaces are C^{p-1}). We may also extend natural operations to the fields as follows. Let τ be a natural element of $\mathcal{L}(A_1 \ldots, A_n; B)$. Given A_1–$, \ldots, A_n$–fields $\kappa_1, \ldots, \kappa_{n'}$ we obtain, applying the construction of the middle of the previous page point wise, a B–field. If, furthermore, $\kappa_1 \ldots, \kappa_n$ are C^{p-1}, then so is this B–field (since a multilinear mapping of Banach spaces, applied to C^{p-1} mappings of Banach spaces, yields a C^{p-1} mapping). Thus, natural operations, applied point wise to C^{p-1} fields, yields C^{p-1} fields.

Example. We have "scalar multiplication", a natural element of $\mathcal{L}(\mathbb{R}, A; A)$. Hence, multiplication of A–fields by scalar fields yields A–fields. The last constructions on pages 47 and 59 are special cases.

Example. The operations "applications". "composition", and "tensor product", applied to C^{p-1} filed, yields, yield C^{p-1} fields.

Example. From the first example and the observation that constant scalar fields (since they arise from natural tensors) are C^{p-1}, we have that set of A–fields form a vector space.

18. Tensor Bundles

Many notions in differential geometry have the feature that they possess two essentially equivalent formulations, one more analytic and algebraic, and the other more geometrical. One might even make a case that some of the interest in the subject arises from this interplay. We have already seen some examples of this phenomenon, e.g., in our discussion of tangent vectors. In the previous section, we treated what might be called the algebraic approach to tensor fields. We now give the geometric one.

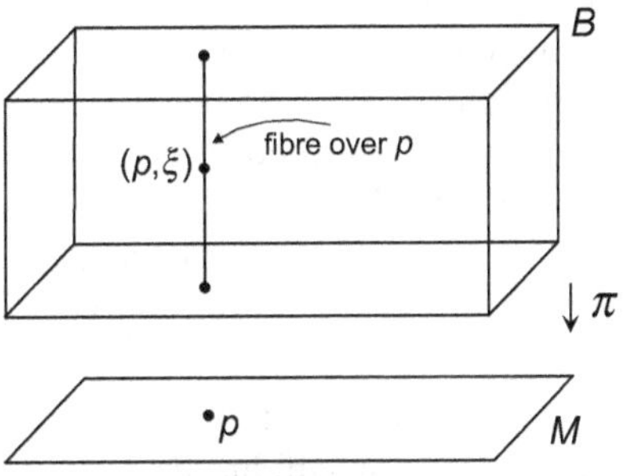

Let M be a C^p ($p \geq 1$) manifold based on Banach space E, and fix a tensor space A over E. Denote by B the set of all pairs, (p, ξ), where p is a point of M, and ξ is an A-tensor at p. Denote by π the mapping from B to M which "ignores the second entry", i.e., with action $\pi(p, \xi) = p$. This B is called the *bundle space*, π the *projection mapping*, and M the *base space*. [In the figure, π is the mapping which "finds the point of M directly under the point of B".] For p a point of M the subset $\pi^{-1}[p]$ (i.e., the set of all elements of B of the form (p, ξ)) of B is called the *fibre* over p. This entire set-up (i.e., the bundle space, projection mapping, base space, fibres) is called the *A-bundle* of M.

We have two goals: to find the structure and properties of the objects defined above, and to describe tensor fields in terms of these objects.

Consider first the bundle space B. We introduce some charts on this set. Let U, ψ be any chart on M. Set $\tilde{U} = \pi^{-1}[U]$, i.e., the union of the fibres over the points of U. Next, let $\tilde{\psi}$ be the mapping from the subset $\tilde{U}$ of B to the Banach space $E \times A$ with the following action: $\tilde{\psi}(p, \xi) = (\psi(p), \alpha)$, where α is component, with respect to the chart U, ψ, of the A-tensor ξ at p. We claim that this $\tilde{U}, \tilde{\psi}$ is an $E \times A$-chart on B. [Proof: Our $\tilde{\psi}$ is one-to-one, for, if $\tilde{\psi}(p, \xi) = \tilde{\psi}(p', \xi')$, then we must have $\psi(p) = \psi(p')$, and hence $p = p'$, and also $\alpha = \alpha'$, and hence $\xi = \xi'$. Further,since $\tilde{\psi}[\tilde{U}] = U \times A$

in $E \times A$, this $\tilde{\psi}[\tilde{U}]$ is open.] Now consider two of these charts on B, $\tilde{U}$, $\tilde{\psi}$ and $\tilde{U}'$,$\tilde{\psi}'$ (from, say, U, ψ and U', ψ' on M). We claim that these charts are C^{p-1}–compatible. [Proof: We have that $\tilde{\psi}[\tilde{U} \cap \tilde{U}'] = \psi[U \cap U'] \times A$, whence this is open in $E \times A$. The mapping $\tilde{\psi}' \cdot \tilde{\psi}^{-1}$ from $\tilde{\psi}[\tilde{U} \cap \tilde{U}']$ to $E \times A$ has action $\tilde{\psi}' \cdot \tilde{\psi}^{-1}(x, \alpha) = (\psi' \cdot \psi^{-1}(x),\ \kappa(x)(\alpha))$, where $\kappa(x)$ is the isomorphism on tensor space A arising from the isomorphism $D(\psi' \cdot \psi^{-1})(x)$ on E. Since composition, application, and insertion into the product are C^{p-1} operations, this mapping $\tilde{\psi}' \cdot \tilde{\psi}^{-1}$ is also C^{p-1}.] Thus, we now have a set B, together with a collection of C^{p-1}–compatible $E \times A$–charts on B. The first, second, and fourth conditions on page 36 are immediate from those conditions for M. We obtain, therefore, a C^{p-1} manifold B based on $E \times A$. [The idea is that "locally, B looks like a product of a small region of M with A".]

Thus, the bundle space B is a manifold. The base space is just M, and it starts out as a manifold. We have already investigated the structure of the fibers: The fibre over p is the set of A–tensors at p, which has the structure of a Banachable space. There remains, therefore, only the projection mapping π. Note that π now is a mapping of C^{p-1} manifolds (from B to M, which can be regarded as a C^{p-1} manifold). It would be natural to guess, therefore, that, this mapping will be C^{p-1}. It is. Proof: Let U, ψ be a chart on M, and $\tilde{U}, \tilde{\psi}$ the corresponding chart on B. Then $\psi \cdot \pi \cdot \tilde{\psi}^{-1}$ is the mapping from the open subset $\tilde{\psi}[\tilde{U}]$ of $E \times A$ to E with action $\psi \cdot \pi \cdot \tilde{\psi}^{-1}(x, \alpha) = x$. But this mapping, "projection onto the first factor", is certainly C^{p-1}. We have shown that π, when "made a mapping of Banach spaces" via certain charts on B and M, yields a C^{p-1} mapping. But, since these "certain charts" cover B and M, the same holds for all charts (since all other charts are compatible with our "certain ones"). Thus, π is a C^{p-1} mapping of manifolds.

To summarize, then, the bundle space is a C^{p-1} manifold, the base space a C^p manifold, and the projection mapping C^{p-1}. The fibres are Banachable spaces.

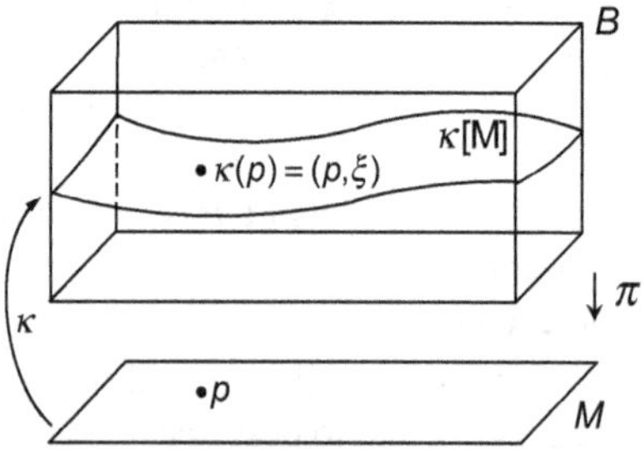

This completes our first goal. We now describe tensor fields in this language. By a *cross section* of our A–bundle we mean a C^{p-1} mapping κ from M to B such that $\pi \cdot \kappa$ is the identity on M. That is to say, we require that, for each p in M, $\kappa(p)$ in B be of the form (p, ξ). Pictorially, one represents a cross-section by drawing $\kappa[M]$ in B as in the figure. We next note that a cross section κ of our A–bundle defines an A–field on M, for each p in M, the "ξ" of $\kappa(p) = (p, \xi)$ is an A–tensor at p. We claim: The A–field so obtained is C^{p-1}. Indeed, choosing chart U, ψ on M, and corresponding chart $\tilde{U}, \tilde{\psi}$ on B, we have since κ is C^{p-1}, that $\tilde{\psi} \cdot \kappa \cdot \psi^{-1}$ is a C^{p-1} mapping from E to $E \times A$. But the action of this mapping sends x in $\psi[U]$ to (x, α), where α is the component of ξ in the chart U, ψ. Hence, the mapping from $\psi[U]$ to A which "evaluates ξ and takes the component" is C^{p-1}. But this is precisely the statement that the A–field ξ is C^{p-1}.

We next claim the converse. Let ξ be a C^{p-1} A–field, and let κ be the mapping from M to B which action $\kappa(p) = (p, \xi(p))$. This κ, we claim, is a cross section. That $\pi \cdot \kappa$ is the identity is obvious, so we need only check C^{p-1}–ness. But, representing the mapping κ in terms of charts as in the previous paragraph, we have, since $\tilde{\psi} \cdot \kappa \cdot \psi^{-1}$ is C^{p-1} (this being what it means for the A–field ξ to be C^{p-1}, that κ is C^{p-1} (for this is what it means for a mapping of manifolds to be C^{p-1}).

We conclude: Cross section of the A–bundle are precisely the same things as (C^{p-1}) A–fields. In this sense, then, we "represent the fields geometrically". We shall see later that these bundles also permit one to draw pictures for various constructions, etc. involving tensor fields.

19. Lie Derivatives

We have now completed our discussion of three broad areas: calculus (Sects 2-8), manifolds (Sects 9-10), and tensor fields (Sects 11-18). The next broad area is that of derivatives (of tensor fields, on manifolds). That is to say, we now wish, for the first time, to make essential use of the symbols "C^p", "C^{p-1}", etc., that we have been carrying along. The basic problem is this: One is given, on a manifold, a certain tensor field, and one wishes to define a new tensor field which, can be interpreted as "the derivative" (in some sense, e.g., with respect to position on the manifold) of the original tensor field.

There is a naive way in which one might attempt to define "derivative" of a tensor field. It is of some interest, first of all, to see why it does not work. Let M be a C^p ($p \geq 2$) manifold based on Banach space E, let A be a tensor space over E, and let α be a C^{p-1} A–field. Fix a point p of M: We try to define "the derivative of α at p" as follows. Let U, ψ be any admissible chart on M, with p in U, and set $x = \psi(p)$. Then the component of our field with respect to this chart is a mapping $\hat{\alpha}$ from the open subset $\psi[U]$ of the Banach space E to the Banach space A. [That is to say, $\hat{\alpha}$ is the following. For q any point of U, $\hat{\alpha}(\psi(q))$, an element of Banach space A, is the component of the A–tensor $\alpha(q)$ at q with respect to the chart U, ψ.] The derivative of this mapping, $D\hat{\alpha}$, is thus a mapping from $\psi[U]$ to the Banach space $\mathcal{L}(E; A)$. Hence, $D\hat{\alpha}(x)$ is an element of $\mathcal{L}(E; A)$. [That is to say, we first "pull the field α over, using a chart, to a mapping from $\psi[U]$ to A". We then "take the derivative of this mapping with respect to the independent variable (point of $\psi[U]$)", and then, finally, we evaluate this derivative at $x = \psi(p)$, the image of p by ψ.] Consider now the pair, $(D\hat{\alpha}(x); U, \psi)$, consisting of a free $\mathcal{L}(E; A)$–tensor and a chart U, ψ. This pair certainly defines one of our equivalence classes, and hence defines a certain $\mathcal{L}(E; A)$–tensor at p. One would like to regard this $\mathcal{L}(E; A)$–tensor at p as "the derivative of the field α on M, evaluated at p". [Note that it is the right sort of object to be so regarded. "Locally near p, M looks like E." Hence, the "derivative of α with respect to position in M" should be a linear mapping from E (which represents "the space of possible directions in which one might move away from p in M") to A. That is, for ξ any tangent vector at p, and end denoting by β the $\mathcal{L}(E;A)$–tensor at p just obtained, then $\beta(\xi)$, an A–tensor at p, would be interpreted as "the derivative of the field α, in the ξ–direction, at p".]

There remains only one little check: chart-independence. If this check should work out properly, then the entire subject "derivatives of tensor fields" would be quite easy (consisting of this and the previous page), we could move on to a new topic, and differential geometry itself would have quite a different character than it has. Of course, it turns out that this check fails. The question, then, is this: Consider a new chart, U', ψ', with p in U'. Then, as above, we obtain a new point, $x' = \psi'(p)$, of $\psi'[U']$ (noting that this x' corresponds to the same p in M), a new mapping $\hat{\alpha}'$ from $\psi'[U']$ to A (noting that this $\hat{\alpha}'$ corresponds to the same A–field α), and a new pair, $(D\hat{\alpha}'(x'); U', \psi')$. Is it true or false that $(D\hat{\alpha}'(x'); U', \psi') \approx (D\hat{\alpha}(x); U, \psi)$, i.e., that we obtain the same $\mathcal{L}(E;A)$–tensor at p via U', ψ' as via U, ψ? To simplify this little calculation, set $A = E$. For q any point of $U \cap U'$, we have, setting $y' = \psi'(q)$ and $y = \psi(q)$, that $\hat{\alpha}'(y') = \tilde{\iota}(\hat{\alpha}(y))$, where $\tilde{\iota} = D(\psi' \cdot \psi^{-1})(y)$ (the formula for how the component of an E–tensor at q changes under change of chart). Setting $\iota = D(\psi' \cdot \psi^{-1})$ we may rewrite this formula as $\hat{\alpha}' \cdot (\psi' \cdot \psi^{-1})(y) = \iota(y)(\hat{\alpha}(y))$. Each side of this equation is a mapping from a certain open subset of $\psi[U]$ (that is where the variable y lives) to the Banach space $\mathcal{L}(E;E)$. Taking the derivative of this equation, using on the left the chain rule and on the right the Leibnitz rule for the derivative of the composition of two y–dependent mappings, we have that $D\hat{\alpha}'(\psi' \cdot \psi^{-1}(y)) \cdot D(\psi' \cdot \psi^{-1})(y) = D\iota(y)(\hat{\alpha}(y)) + \iota(y)(D\hat{\alpha}(y))$. Now set $y = x$ (so $\psi' \cdot \psi^{-1}(y) = x'$), and use the definition of ι for the second expression on the left to obtain $D\hat{\alpha}'(x') \cdot \iota(x) = D\iota(x)(\hat{\alpha}(x)) + \iota(x)(D\hat{\alpha}(x))$. Finally, applying $\iota(x)^{-1}$ to both sides, we obtain our desired equation: $\iota(x)^{-1} D\hat{\alpha}'(x')\iota(x) = \iota(x)^{-1} D\iota(x)(\hat{\alpha}(x)) + D\hat{\alpha}(x)$. Now, this last formula is just true. What is it that we want to show? It is that $(D\hat{\alpha}'(x'); U', \psi') \approx (D\hat{\alpha}(x); U, \psi)$. Using the action of ι on $\mathcal{L}(E;E)$, this amount precisely to showing that $\iota(x)^{-1} D\hat{\alpha}'(x')\iota(x) = D\hat{\alpha}(x)$. Comparing what we have and what we want, we see that there is an extra term in the former $(\iota(x)^{-1} D\hat{\alpha}(x)(\alpha(x)))$ – a term which will not in general be zero. We conclude that our check fails. In general, we shall not obtain, by the prescription of the previous page, an $\mathcal{L}(E : A)$– tensor at p which is independent of the chart used in that prescription.

All the mapping above tend to obscure what is basically a simple idea. The problem is that the "change in component under change in chart" mapping ι in general will depend on y. To pass from our A–field to our proposed $\mathcal{L}(E;A)$–field, we must take a derivative. Thus, although the component of α behaves properly (i.e., algebraically in ι) under chart-change, the "derivative of component" picks up an extra term involving the derivative of ι. This extra term prevents the "derivative of component" from behaving properly under chart-change, i.e., prevents "derivative of component" from representing the component of some other field on M.

The problem of taking derivatives of tensor fields amounts essentially to the problem of finding various ways to get rid of the "extra term" in the little calculation above. It turns out that there are at least three such ways, where the corresponding derivatives are called Lie derivatives, exterior derivatives, and derivative operators. Each of these types of derivative has its own advantages and disadvantages. In general terms, the advantage of each is that it "looks and acts like a derivative operation", and the disadvantage that some extra structure or restriction of the action has been employed to eliminate the extra term we found above. It is our proposal now to study these three kinds of derivatives, beginning with the Lie derivative.

It will be necessary, below, to make use of the following mapping. Let E be any Banach space, and A any tensor space over A. Denote by $\mathcal{L}_{\text{inv}}(E;E)$ the set of all invertible elements of $\mathcal{L}(E;E)$ (an open subset of the Banach space $\mathcal{L}(E;E)$, and hence a C^∞ manifold based on $\mathcal{L}(E;E)$). Next, let φ denote the following mapping from $\mathcal{L}_{\text{inv}}(E;E)\times A$ to A: For ι in $\mathcal{L}_{\text{inv}}(E;E)$ and α in A, $\varphi(\iota,\alpha)$ is the element of A obtained by first extending the isomorphism ι on E to A (Sect. 15), and then applying the resulting isomorphism from A to A to the element α of A. Thus, for example, we have $\varphi(\iota\cdot\iota',\alpha) = \varphi(\iota,\varphi(\iota',\alpha)), \varphi(\iota,\alpha+\alpha') = \varphi(\iota,\alpha)+\varphi(\iota,\alpha')$, and $\varphi(\iota,\alpha) = \alpha$ if α, is natural. This φ is a C^∞ mapping of manifolds. Hence, fixing α, the mapping φ_α from $\mathcal{L}_{\text{inv}}(E;E)$ to A with action $\varphi_\alpha(\iota) = \varphi(\iota,\alpha)$ is also C^∞. The derivative of this mapping, $D\varphi_\alpha$, is therefore a mapping from $\mathcal{L}_{\text{inv}}(E;E)$ to $\mathcal{L}(\mathcal{L}(E;E);A)$. Applying to the identity I (certainly an element of $\mathcal{L}_{\text{inv}}(E;E)$), we have $D\,\varphi_\alpha(I)$, an element of $\mathcal{L}(\mathcal{L}(E;E);A)$. Finally, let κ be the mapping from $\mathcal{L}(E;E)\times A)$ to A with action $\kappa(\xi,\alpha) = D\varphi_\alpha(I)(\xi)$. This κ is, clearly, bilinear, i.e., is an element of $\mathcal{L}(\mathcal{L}(E;E),A;A)$.

Example. Let $A = R$. Then φ has action $\varphi(\iota,r) = r$ (action of our isomorphisms on the reals), whence φ_r is the constant mapping (i.e., $\varphi_r(\iota) = r$ for every ι. Thus, $D\varphi_r = 0$, whence $\kappa = 0$.

Example. Let $E = E$. Then φ has action $\varphi(\iota,x) = \iota(x)$. Hence (since φ_x is a linear mapping), $D\varphi_x(\iota,\xi) = \xi(x)$, for ξ in $\mathcal{L}(E;E)$. Therefore, $\kappa(\xi,x) = \xi(x)$.

Example. Let $A = \mathcal{L}(E;E)$. Then φ has action $\varphi(\iota,\alpha) = \iota\cdot\alpha\cdot\iota^{-1}$, whence $D\varphi_\alpha(\iota)(\xi) = \xi\cdot\alpha\cdot\iota^{-1} - \iota\cdot\alpha\cdot\iota^{-1}\cdot\xi\cdot\iota^{-1}$. Therefore, $\kappa(\xi,\alpha) = \xi\cdot\alpha - \alpha\cdot\xi$.

Now fix, once and for all, a C^p ($p \geq 2$) manifold M based on Banach space E. The Lie derivative will be essentially a "generalized directional derivative" (generalized from action on scalar fields to all tensor fields on M). We shall need some direction in which to take the derivative. Hence, let ξ be a C^{p-1} tangent vector field (E–field) on M. Finally, let A be any tensor space over E, and let α be a C^{p-1} A–field on M. The idea is to define, from ξ and α, a new A–field on E.

Let U,ψ be a chart on M, and let η and β be the components of ξ and α, respectively (so η is a C^{p-1} mapping from $\psi[U]$ to E, while β is a C^{p-1}

mapping from $\psi[U]$ to A). Consider now the mapping ω from $\psi[U]$ to A with action $\omega(x) = D\beta(x)(\eta(x)) - \kappa(D\eta(x), \beta(x))$ (noting that all this is well-defined: $D\beta(x)$ is an element of $\mathcal{L}(E;A)$, whence $D\beta(x)\,(\eta(x))$ is in A : $D\eta(x)$ is an element of $\mathcal{L}(E;E)$ and $\beta(x)$ an element of A, whence $\kappa(D\eta(x), \beta(x))$ is an element of A). Thus, we have so far written down a mapping ω from $\psi[U]$ to A. This ω is our candidate for the component of a certain A–field on M, namely, the component with respect to the chart U, ψ. To verify that this ω actually leads to an A–field, we must check to see how ω changes when the chart is changed. To this end, let $U'.\psi'$ be a different chart. Then, denoting by η' and β' the components of ξ and α, respectively, with respect to this chart, we have that $\beta'(x') = \varphi(\iota(x), \beta(x))$ and $\eta'(x') = \iota(x)(\eta(x))$, where $x' = \psi' \cdot \psi^{-1}(x)$ (the statement that the point x' of $\psi'[U']$ defines the same point of M as the point x of $\psi[U]$) and $\iota = D(\psi' \cdot \psi^{-1})$. These two formulae, then, are just those for the behaviour of the component under chart-change. Rewriting the first equation in the form $\beta' \cdot (\psi' \cdot \psi^{-1})(x) = \varphi_{\beta(x)}(\iota(x))$, taking the derivative (with respect to x), and applying to an arbitrary element v of E, we obtain $D\beta'(x')\iota(x)(v) = \varphi_D\beta(x)(v)(\iota(x)) + \kappa((D\iota(x)(v))\iota^{-1}(x),\ \varphi(\iota(x)\beta(x))$ where we used the chain rule on the left, the formula for the derivative of the application of a bilinear mapping to x–dependent vectors on the right, and the definition of κ in the last term. Now set $v = \eta(x)$ in this equation, to obtain $D\beta'(x')(\eta'(x')) = \varphi(\iota(x), D\beta(x)(\eta(x))) + \kappa(D\iota(x)(\eta(x))\iota^{-1}(x),\ \varphi(\iota(x), \beta(x)))$. Next, take the derivative of the formula for the component change of ξ to obtain $D\eta'(x')\iota(x)(v) = (D\iota(x)(v))(\eta(x)) + \iota(x)(D\eta(x)(v))$, where v is an arbitrary element of E. Using the fact that mixed partials commute on the first term on the right, we have that this term is $(D\iota(x)(\eta(x))(v)$. Inserting this into the expression above, and using the fact that v is arbitrary, we have $D\eta'(x')\iota(x) = D\iota(x)(\eta(x)) + \iota(x)(D\eta(x))$. That is to say, $D\iota(x)(\eta(x))\iota^{-1}(x) = D\eta'(x') - \iota(x)(D\eta(x))\iota^{-1}(x)$. Substituting this into the formula on line seventeen above, we obtain $D\beta'(x')(\eta'(x')) = \varphi(\iota(x), D\beta(x)(\eta(x))) + \kappa(D\eta'(x'), \varphi(\iota(x), \beta(x)) - \kappa(\iota(x)D\eta(x)\iota^{-1}(x), \varphi(\iota(x), \beta(x))$. But the second term on the right is just $\kappa(D\eta'(x'), \beta'(x')))$. Substituting, we obtain finally, $D\beta'(x')(\eta'(x')) - \kappa(D\eta'(x'), \beta'(x')) = \varphi(\iota(x), D\beta(x)(\eta(x)) - \kappa(D\eta(x), \beta(x))$. But this formula is precisely the statement that $\omega'(x') = \varphi(\iota(x), \omega(x))$, i.e., the statement that, for each x, $(\omega'(x'); U', \psi') \approx (\omega(x); U, \psi)$. We conclude, therefore, that our ω indeed has the proper behaviour under changes in chart.

We now define the *Lie derivative*, $\mathcal{L}_\xi\alpha$, of the A–field α in the ξ–direction by the formula above. That is to say, letting η, β, and ω be the components of ξ, α, and $\mathcal{L}_\xi$ respectively, with respect to a chart, we set $\omega(x) = D\beta(x)(\eta(x)) - \kappa(D\eta(x), \beta(x))$.

The calculation above is rather messy because of its generality, and because there are so many mappings around. It is easier to see what is going on by looking at examples.

Example. Set $A = \mathbb{R}$. Thus, we wish to take the Lie derivative of scalar field α in the ξ–direction. In this case, β is a mapping from $\psi[U]$ to $\mathbb{R}$, and κ is zero. Hence, the formula above becomes $\omega(x) = D\beta(x)(\eta(x))$. But the right side will be recognized as the formula for the directional derivative of α in the direction of the tangent vector $\xi(p)$ at $p = \psi^{-1}(x)$. Thus, the Lie derivative of a scalar field is its directional derivative.

Example. Set $A = E$. We wish to take the Lie derivative of the tangent vector field α in the ξ–direction. Now κ has action $\kappa(\zeta, \nu) = \zeta(\nu)$, whence our formula above becomes $\omega(x) = D\beta(x)(\eta(x)) - D\eta(x)(\beta(x))$. This ω,then, is the component of $\mathcal{L}_\xi\alpha$. Note in particular from this formula that $\mathcal{L}_\xi\alpha = -\mathcal{L}_\alpha\xi$.

Example. Set $A = \mathcal{L}(E; E)$. Then κ has action $\kappa(\zeta, \mu) = \zeta \cdot \mu - \mu \cdot \zeta$. Hence, the formula for the component of $\mathcal{L}_\xi\alpha$ is $\omega(x) = D\beta(x)(\eta(x)) - D\eta(x) \cdot \beta(x) + \beta(x) \cdot D\eta(x)$.

It is a good exercise to verify explicitly that the $\mathcal{L}_\xi\alpha$ of each example above has the proper behavior under changes of chart.

We now have a thing, the Lie derivative, with a rather complicated definition. We deal with this situation in the usual way: We attempt to find a list, of reasonable length, of properties of the Lie derivative, with the goal of using in practice the properties on the list rather than the original definition. Such a list follows.

Property 1. The Lie derivative of a C^{p-1} A–field in the ξ–direction (where ξ is a C^{p-1} E–field) is a C^{p-2} A–field. More generally, if the A–field is C^q $(1 \le q \le (p-1)$, or, for $A = \mathbb{R}$, $1 \le q \le p)$, and ξ is $C^{q'}$ $(1 \le q' \le (p-1))$, then $\mathcal{L}_\xi\alpha$ is $C^{q''}$, where q'' is the minimum of $q-1$ and $q'-1$. [This fact is immediate from the formula, and the fact that composition, application, etc. are C^∞ operations.]

Property 2. For α an R–field, $\mathcal{L}_\xi\alpha$ is the directional derivative of α in the ξ–direction. (Example, page 92.)

Property 3. For α an R–field, and ξ and τ E–fields, we have $\mathcal{L}_\xi(\mathcal{L}_\tau\alpha) - \mathcal{L}_\tau(\mathcal{L}_\xi\alpha) = \mathcal{L}_{\mathcal{L}_\xi\tau}\alpha$. Proof: Choose a chart, and let the components of ξ, τ and α be η, σ, and β, respectively. Then the component of $\mathcal{L}_\tau\alpha$ sends x to $\omega(x) = D\beta(x)(\sigma(x))$, whence the component of $\mathcal{L}_\xi(\mathcal{L}_\tau\alpha)$ sends x to $D\omega(x)(\eta(x)) = DD\beta(x)(\sigma(x), \eta(x)) + D\beta(x)[D\sigma(x)(\eta(x))]$. Reversing the roles of ξ and τ, the component of $\mathcal{L}_\xi\mathcal{L}_\tau\alpha - \mathcal{L}_\tau\mathcal{L}_\xi\alpha$ sends x to $DD\beta(x)(\sigma(x), \eta(x)) + D\beta(x)[D\sigma(x)(\eta(x))) - DD\beta(x)(\eta(x), \sigma(x)) - D\beta(x)[D\eta(x)(\sigma(x))]$. But the first and third terms cancel, since mixed partials are symmetric, and so we obtain $D\beta(x)[D\sigma(x)(\eta(x)) - D\eta(x)(\sigma(x))]$. But this is precisely the component of $\mathcal{L}_{\mathcal{L}_\xi\tau}\alpha$. In fact, it is also true that $\mathcal{L}_\xi(\mathcal{L}_\tau\alpha) - \mathcal{L}_\tau(\mathcal{L}_\xi\alpha) = \mathcal{L}_{\mathcal{L}_\xi\tau}\alpha$ for any A–field α. The proof is essentially the same as that above: Write down both sides in terms of a chart.

Property 4. For α and α' A–fields, $\mathcal{L}_\xi(\alpha + \alpha') = \mathcal{L}_\xi\alpha = \mathcal{L}_\xi\alpha'$. This is immediate from the defining formula and the fact that κ is bilinear.

Property 5. For τ a $\mathcal{L}(A; B)$–field, and α an A–field, we have $\mathcal{L}_\xi(\tau(\alpha)) =$

$(\mathcal{L}_\xi\tau)(\alpha) + \tau(\mathcal{L}_\xi\alpha)$. Proof: For $\hat{\tau}$ in $\mathcal{L}(A;B)$ and $\hat{\alpha}$ in A we have, taking the derivative of the action of ι on $\mathcal{L}(A;B)$, that $\kappa(\zeta,\hat{\tau}(\hat{\alpha}) = \kappa(\zeta\hat{\tau})(\hat{\alpha}) + \hat{\tau}(\kappa(\zeta,\hat{\alpha}))$ for every ζ in $\mathcal{L}(E;E)$. Denote by $\hat{\tau}$ and $\hat{\alpha}$ the components of τ and α, respectively. Then we have that the action of the component of $\mathcal{L}_\xi(\tau(\alpha))$ is $\omega(x) = D(\hat{\tau}(\hat{\alpha}))(x)\eta(x) - \kappa(D\eta(x),\hat{\tau}(x)(\hat{\alpha}(x)))$. Expand the first term on the right using the Leibnitz rule (for derivatives of mappings of Banach spaces), and use the fact above in the second. The result is the equation claimed, written in terms of our chart. By a similar argument one obtains: For τ a $\mathcal{L}(A_1,\ldots,A_n;B)$–field, and $\alpha_1,\ldots,\alpha_n A_1-,\ldots,A_n$–fields, $\mathcal{L}_\xi(\tau(\alpha_1,\ldots,\alpha_n)) = (\mathcal{L}_\xi\tau)(\alpha_1,\ldots,\alpha_n)+\tau(\mathcal{L}_\xi\alpha_1,\alpha_2,\ldots,\alpha_n)+\ldots+\tau(\alpha_1,\ldots,\mathcal{L}_\xi\alpha_n)$. Thus, property 5 is the Leibnitz rule for Lie derivatives.

Property 6. For α any natural A–field, $\mathcal{L}_\xi\alpha = 0$. Proof: For α natural, and $\hat{\alpha}$ the component of α, we have that the mapping $\hat{\alpha}$ is constant (for $\hat{\alpha}(x)$ is this natural tensor for each x). Furthermore, $\kappa(\zeta,\hat{\alpha}(x)) = 0$ for any ζ in $\mathcal{L}(E;E)$. Hence, for the component of $\mathcal{L}_\xi\alpha$, we have $D\hat{\alpha}(x)(\hat{\xi}(x)) - \kappa(D\hat{\xi}(x),\hat{\alpha}(x)) = 0$ (the first term vanishing since it is a derivative of a constant map). Thus, since the component of $\mathcal{L}_\xi\alpha$ in every chart vanishes, we have $\mathcal{L}_\xi\alpha = 0$.

Property 7. The Leibnitz rule is satisfied for any natural tensor operation. That is, for τ natural in $\mathcal{L}(A_1,\ldots,A_n;B)$, and $\alpha_1,\ldots,\alpha_n A_1-,\ldots,A_n$–tensors, $\mathcal{L}_\xi(\tau(\alpha_1,\ldots,\alpha_n)) = \tau(\mathcal{L}_\xi\alpha_1,\ldots,\alpha_n) + \ldots + \tau(\alpha_1,\ldots,\mathcal{L}_\xi\alpha_n)$. Proof: Immediate from properties 5 and 6.

Properties 4 and 7 summarize the dependence of "$\mathcal{L}_\zeta\alpha$" on α. We next consider its dependence on ξ. To this end, first note that, for any tensor space A, the element κ of $\mathcal{L}(\mathcal{L}(E;E),A;A)$ is natural. Hence we obtain, on our manifold, a corresponding natural $\mathcal{L}(\mathcal{L}(E;E),A;A)$–field, which we also denote κ.

Property 8. For α and A–field, and ξ and ξ' E–fields, we have $\mathcal{L}_{\xi+\xi'}\alpha = \mathcal{L}_\xi\alpha + \mathcal{L}_{\xi'}\alpha$. Immediate.

Property 9. Let α be an A–field, ξ an E–field, and f an $\mathbb{R}$–field. Denote by ζ the $\mathcal{L}(E;E)$–field with action $\zeta(v) = \mathcal{L}_v f)\xi$ for every E–field v. Then $\mathcal{L}_{(f\xi}\alpha = f(\mathcal{L}_\xi\alpha) + \kappa(\zeta,\alpha)$. Proof: The component of $\mathcal{L}_{(f\xi)}\alpha$ has action $\omega(x) = D\hat{\alpha}(x)(f(x)\hat{\xi}(x)) - \kappa(D(f\hat{\xi})(x),\hat{\alpha}(x))$. Expanding the second term on the right using the Leibnitz rule for derivatives, this becomes $\omega(x) = D\hat{\alpha}(x)(f(x)\hat{\xi}(x)) - \kappa(f(x)D\hat{\xi}(x),\hat{\alpha}(x)) - \kappa(\zeta(x),\hat{\alpha}(x))$. Each of the first two terms on the right is now linear in $\xi(x)$, whence the $f(x)$, a number, can be pulled outside. The result is precisely the component form of the claimed equally.

This completes our list of properties.

Example. For α and A–field, and f and $\mathbb{R}$–fields, $\mathcal{L}_\xi(f\alpha) = (\mathcal{L}_\xi f)\alpha + f(\mathcal{L}_\xi\alpha)$. Property 7.

Example. For α an A–field and β a B–field, $\mathcal{L}_\xi(\alpha\otimes\beta) = (\mathcal{L}_\xi\alpha)\otimes\beta+\alpha\otimes(\mathcal{L}_\xi\beta)$. Property 7.

Example. For ξ and η E–fields, and f and R–field, $\mathcal{L}_{(f\xi)}\eta = f(\mathcal{L}_\xi\eta) +$

$(\mathcal{L}_\eta f)\xi$. Property 9, or else properties 3 and 7.

It is common in the finite-dimensional case to define the Lie derivative, not by the chart-formula that we have used here, but rather by some combination of the properties above. The idea is to show that there is one and only one "Lie derivative operation" satisfying certain properties, and then establish our chart-formula as a theorem. In more detail, one uses property 2 to get the action of the Lie derivative on scalar fields, then property 3 to get the action on vector fields, and finally various special cases of property 7 to get the action on other tensor fields. One could certainly carry out a similar program in the infinite-dimensional case but, unfortunately, a number of difficulties intervene to make this somewhat awkward. We here just mention what these difficulties are and how they might be surmounted. First, one's manifold may not admit a reasonable number of fields (as we have seen in an example, for scalar fields). One is thus forced, apparently, to introduce a "Lie derivative operation" locally, in small open regions, and then piece together these regions to get the operation over the entire manifold. Second, one must apparently impose all the differentiability assignments (property 1) on one's Lie derivative operation (including the exception for scalar fields). If one, for example, ignored the fact that scalar fields can be C^p (whereas others must be C^{p-1}), then one would not, by property 3, obtain a C^{p-2} vector field for the Lie derivative of a vector field. [The same problem arises in the finite-dimensional case. However, in finite dimensions one often works in C^∞, because it seems to turn out in practice that nothing is lost by such a restriction. Property 1 then simplifies somewhat. However, it is not so clear that "everything C^∞" is a reasonable condition in infinite-dimensions.] Third, one has to go to charts anyway, in the use of property 3. From this property, one will indeed find $(\mathcal{L}_\xi\eta)(f)$, once one knows Lie derivatives of scalar field. One must, however, then show that this is the directional derivative of f by some vector field (to be identify with $\mathcal{L}_\xi\eta$). This demonstration requires, apparently, charts. Fourth, there are apparently some problems with tensor products. We can of course define $\mathcal{L}_\xi(\alpha \otimes \beta)$ by the second example on the previous page. What, however, is the formula for $\mathcal{L}_\xi\tau$ when τ is an arbitrary $A \otimes B$–tensor? It is obvious to me that every such τ can be written in the form $\tau = \alpha_1 \otimes \beta_1 + \alpha_2 \otimes \beta_2 + \ldots$ (so that we could set $\mathcal{L}_\xi\tau = (\mathcal{L}_\xi\alpha_1) \otimes \beta_1 + \alpha_1 \otimes (\mathcal{L}_\xi\beta_2) + \ldots$). Furthermore, even if this were true, we would still have to show that the second sum converges (in the tensor product) provided that the first one does. Smoothness would be a further problem. Finally, I am aware of no way, in this program, to show directly (i.e., without going back to charts) that the Lie derivative of every natural tensor vanishes. [This problem does not arise in the finite-dimensional case, since there one has the complete list of natural tensors, and simply checks them one at a time.]

It is our claim that all the various properties of the Lie derivative give expression to the idea that "the Lie derivative is a generalized directional derivative". The nice thing about the Lie derivative is that it can be applied to any tensor field, and it has a large number of properties. The disadvantage is that it requires some particular choice of a vector field, ξ, along which the Lie derivative is taken.

20. Integral Curves

The definition of the previous section may be called the algebraic-analytic approach to Lie derivatives. There is also a more geometrical approach to the same subject. It is our intention to discuss this alternative viewpoint. We require, as a prerequisite, the notion of an integral curve, and some properties of these. This general subject – integral curves – has numerous applications, extending far beyond just Lie derivatives, one of which we indicate briefly.

Fix, once and for all, a C^p ($p \geq 2$) manifolds M based on Banach space SE, and a C^{p-1} E–field ξ on M. An *integral curve* of ξ consists of an open interval, (a, b), of the reals, together with a C^p mapping γ from (a, b) (regarded as a manifold) to M, such that the following property is satisfied: For each number r in (a, b), the tangent vector to the curve γ at r is precisely $\xi(\gamma(r))$, i.e., the value of the field ξ at the point $\gamma(r)$ of M. Intuitively, an integral curve is "always moving along in M tangentially to ξ".

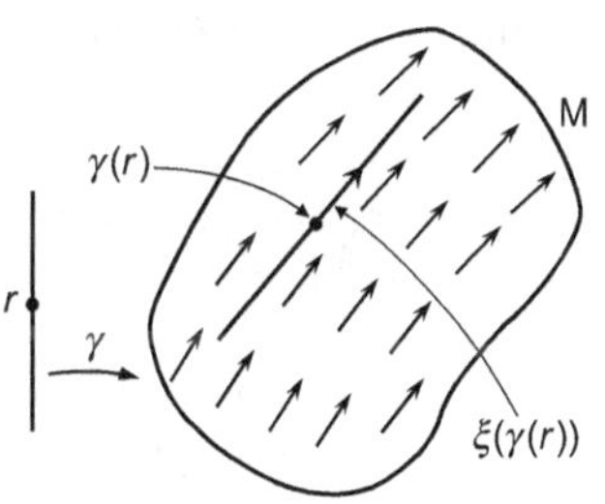

Example. Let (a, b),γ be an integral curve, and let s be any real number. Then $(a + s, b + s)$, $\tilde{\gamma}$ is also an integral curve,where $\tilde{\gamma}$ has action $\tilde{\gamma}(r) = \gamma(r - s)$. These two integral curves are said to be related by *reparameterization*.

Example. Let $M = E$. Fix any ξ_0 in E, and let ξ be that vector field on M whose component in this chart is ξ_0 for all x. Then, in terms of this chart, a typical integral curve is given by (a, b), γ, where $\gamma(r) = x_0 + r\xi_0$, where x_0 is any fixed vector in E.

It is our goal to decide whether or not integral curves exist, and how unique they are. It is convenient, for this discussion, to have available three definitions. For (a, b), γ an integral curve, with 0 in (a, b), the point $\gamma(0)$ of M is called the *initial point* of the curve. Clearly, e.g., a reparameterization serves merely to shift the initial point along the integral curve. Next, let (a, b), and (a', b'), γ' be two integral curves. The second is said to be an *extension* of the first if i) $a' \leq a$ and $b \leq b'$ (i.e., (a', b') includes (a, b)), and ii) $\gamma(r) = \gamma'(r)$ for r in (a, b) (i.e., wherever both curves are defined). Thus, an extension of an integral curve merely "makes the curve longer, on one or both ends". Finally, an integral curve, the only extension of which is the curve itself, is called *maximal.* [So, as one might expect, "is an extension of" is a partial ordering on integral curves.] Thus, for example, an integral curve is maximal if and only if every reparametrization is; the curve of the second example of page 97 is maximal provided $a = -\infty$ and $b = \infty$.

Clearly, integral curves are only going to be unique up to reparametrizations and extensions. The theorem on existence and uniqueness of integral curves is the simplest and strongest one could expect in light of this observation.

Theorem. Let M be a C^p $(p \geq 2)$ manifold based on E, ξ a C^{p-1} E–field, and p a point of M. Then there exists one and only one maximal integral curve γ of ξ with initial point p.

Proof: We first make the following observation. Let U, ψ, be a chart on M, let $O = \psi[U]$, let $\hat{\xi}$ (a mapping from O to E) be the component of ξ, and let $\hat{\gamma} = \psi \cdot \gamma$ (a curve on O). Then the statement that γ is an integral curve becomes, in terms of this chart, the equation $D\hat{\gamma} = \hat{\xi} \cdot \hat{\gamma}$.

Uniqueness. Let O and $\hat{\xi}$ be as above, and let c and d be numbers such that $c = \text{lub}\,|\hat{\xi}(x)|$ and $d = \text{lub}\,|\mathrm{D}\hat{\xi}(\mathrm{x})|$. Let (a, b) be an open interval with $a < 0 < b$, and let $\hat{\gamma}$ and $\hat{\gamma}'$ be mappings from (a, b) to O satisfying $D\hat{\gamma} = \hat{\xi} \cdot \hat{\gamma}$ $D\hat{\gamma}' = \hat{\xi} \cdot \hat{\gamma}'$, and $\hat{\gamma}(\beta) = \hat{\gamma}'(0) = x_0$. Finally, let τ have action $\tau(x) = \hat{\gamma}(x) - \hat{\gamma}'(x)$. We have $D\tau(r) = D\hat{\gamma}(r) - D\hat{\gamma}'(r) = hat\xi(\hat{\gamma}(r)) - \hat{\xi}(\hat{\gamma}'(r)) \leq \text{lub}\,|\mathrm{D}\hat{\xi}|\,|\hat{\gamma}(\mathrm{r}) - \hat{\gamma}'(\mathrm{r})| = \mathrm{d}\,|\tau(\mathrm{r})|$, were we used the mean value theorem in the third step and definitions in the fourth. Hence, $|\tau(r)| = |\tau(r) - \tau(0)| \leq \underset{0 \leq r' \leq r}{\text{lub}}\,|D\tau(r')|\,|r| \leq d \underset{0 \leq r' \leq r}{\text{lub}}\,|\tau(r')|\,|r|$. Now choose positive r_0 such that $r_0 d \leq 1/2$. Then, taking the lub of the formula just obtained, we have $\underset{0 \leq r \leq r_o}{\text{lub}}\,|\tau(r)| \leq d r_0 \text{lub}\,|\tau(\mathrm{r})| \leq 1/2 \underset{0 \leq \mathrm{r} \leq \mathrm{r}_0}{\text{lub}}\,|\tau(\mathrm{r})|$. But this is possible only if $\text{lub}\,|\tau(\mathrm{r})| = 0$ i.e., if $\tau(r) = 0$, for every r in $[), r_0]$. That is, $\hat{\gamma}(r) = \hat{\gamma}'(r)$ for $0 \leq r \leq r_0$.

Now consider two maximal integral curves γ and γ' with $\gamma(0) = \gamma'(0) = p$, as in the theorem. Let s be the largest.

Now consider two maximal integral curves, $(a, b), \gamma$ and $(a', b'), \gamma'$, with $\gamma(0) = \gamma'(0) = p$, as in the theorem. Let s be the largest number such that $s \leq b$, $s \leq b'$, and $\gamma(r) = \gamma'(r)$ whenever $0 \leq r < s$. We claim that the assumption $s < b$ leads to a contradiction. Indeed, we must have, under this assumption, $s < b'$, for otherwise the integral curve γ would lead to an extension of γ', contradicting maximality of γ'. Furthermore, by continuity, we must have $\gamma(s) = \gamma'(s)$. Choosing a chart U, ψ in M, with $\gamma(s) = \gamma'(s)$ in U, we have, by the result of the previous paragraph, that $\gamma(r) = \gamma'(r)$ also for $s \leq r \leq s + r_0$, for some positive r_0. But this violates the definition of s. This contradiction establishes that $s = b$, and, similarly $s = b'$. That is to say, we have $b = b'$ and $\gamma(r) = \gamma'(r)$ for $0 \leq r < b$. In a similar way, "working in the other direction", we have that $a = a'$, and $\gamma(r) = \gamma'(r)$ for $a < r \leq 0$. That is to say, our two maximal integral curves are identical.

Existence. Let O be an open subset of Banach space E, $\hat{\xi}$ a C^{p-1} mapping from O to E, and let c and d be numbers such that $c = \text{lub}\,|\xi(\text{x})|$ and $d = \text{lub}\,|\text{D}\xi(\text{x})|$. Finally, let p and $\tilde{p}$ be two points of O, let ϵ be a sufficiently small positive number, and let α be a number with $0 \leq \alpha \leq 1$. Set $s = |p - \tilde{p}|$. We now associate, with each of p and $\tilde{p}$, a new point of O as follows. With p, associate the point $p + \epsilon\hat{\xi}(p)$. With $\tilde{p}$, associate the point obtained by first finding $\tilde{p} + \alpha\epsilon\hat{\xi}(\tilde{p})$, and then $\tilde{p} + \alpha\epsilon\hat{\xi}(\tilde{p}) + (1 - \alpha)\epsilon\hat{\xi}(\tilde{p} + \alpha\epsilon\hat{\xi}(\tilde{p})$. [That is, the first point is obtained by "going amount ϵ along the direction of $\hat{\xi}(p)$ from p"; the second point by "first going amount $\alpha\epsilon$ along the direction of $\hat{\xi}(\tilde{p})$ from $\tilde{p}$ to obtain $\underset{\sim}{p} = \tilde{p} + \alpha\epsilon\hat{\xi}(\tilde{p})$, and then going amount $(1 - \alpha)\epsilon$ along the direction of $\hat{\xi}(\underset{\sim}{p})$ from $\underset{\sim}{p}$".] We wish to find the "distance" s' between the two points so associated. We have $s' = |p + \epsilon\hat{\xi}(p) - \tilde{p} - \alpha\epsilon\hat{\xi}(\tilde{p}) - (1 - \alpha)\epsilon\hat{\xi}(\tilde{p} + \alpha\epsilon\hat{\xi}(\tilde{p}))| = |p - \tilde{p} + \epsilon\hat{\xi}(p) - \epsilon\hat{\xi}(\tilde{p}) + (1 - \alpha)\epsilon\hat{\xi}(\tilde{p}) - (1 - \alpha)\epsilon\hat{\xi}(\tilde{p} + \alpha\epsilon\hat{\xi}(\tilde{p}))| \leq s + \epsilon\,|\hat{\xi}(p) - \hat{\xi}(\tilde{p})| + |(1 - \alpha)\epsilon\hat{\xi}(\tilde{p}) - (1 - \alpha)\epsilon\hat{\xi}(\tilde{p} + \alpha\epsilon\hat{\xi}(\tilde{p}))| \leq s + \epsilon\, d\, s + (1 - \alpha)\epsilon\,|\hat{\xi}(\tilde{p}) - \hat{\xi}(\tilde{p} + \alpha\epsilon\hat{\xi}(\tilde{p}))| \leq s(1 + \epsilon\, d) + (1 - \alpha)\epsilon(d\,|\alpha\epsilon\hat{\xi}(\tilde{p})|) \leq s(1 + \epsilon\, d) + (1 - \alpha)\epsilon\, d\, \alpha\epsilon\, c \leq s(1 + \epsilon\, d) + \epsilon^2 cd/4$, where the first step is definition of s', the second results from rearranging terms, the third is the triangle inequality, the fourth results from using the mean value theorem on the second term, the fifth results from using the mean value theorem on the third term, the sixth uses the definition of c, and the seventh uses the fact that $\alpha(1 - \alpha) \leq 1/4$. We shall use this formula, $s' \leq s(1 + \epsilon\, d) + \epsilon^2 cd/4$, several times in what follows.

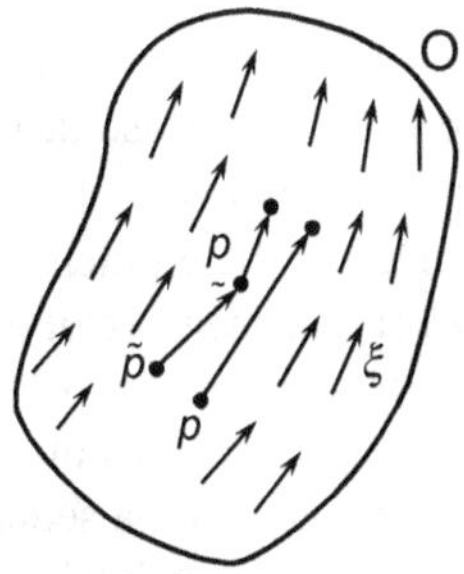

Next, fix a point p of O and a sufficiently small positive number $\underline{r}$. Let $r_0, \ldots, r_n$ be numbers with $0 = r_0 < r_1 < \ldots < r_n = \underline{r}$, and set $\epsilon = \max|\text{r}_{\text{i}+1} - \text{r}_\text{i}|$. We construct, from this set-up, a point of O as follows: First find $p_1 = p + r_1\hat{\xi}(p)$, then $p_2 = p_1 + (r_2 - r_1)\hat{\xi}(p_1)$, then, $p_3 = p_2 + (r_3 - r_2)\hat{\xi}(p_2)$, etc. Continue, until one finds p_n. Now let there be given numbers $\alpha_1, \ldots, \alpha_n$, all between zero and one. Then we can find a second point of O by using the α's to subdivide as above. That is, set $\tilde{p}_1 = p + \alpha_1 r_1\hat{\xi}(p) + (1 - \alpha_1)r_1\hat{\xi}(p + \alpha_1 r_1\hat{\xi}(p))$, then $\tilde{p}_2 = \alpha_2 r_2\hat{\xi}(\tilde{p}_1) + (1 - \alpha_2)(r_2\hat{\xi}(\tilde{p}_1 + \alpha_2 r_2\hat{\xi}(\tilde{p}_1))$, etc. Set $s_i = |p_i - \tilde{p}_i|$, and write μ for $1 + \epsilon d$ and ν for $\epsilon^2 cd/4$. Then, by our formula above, $s_1 \leq \nu$. Applying the same for-

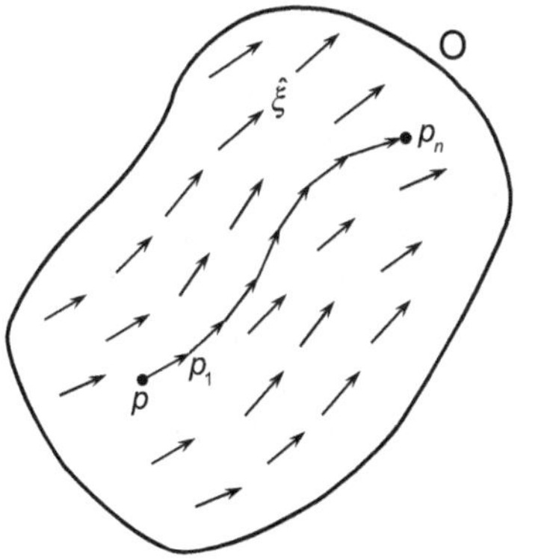

mula for the second step, we have $s_2 \le \mu s_1 + \nu \le \nu(1+\mu)$. For the third, $s_3 \le \mu s_2 + \nu \le \nu(1+\mu+\mu^2)$. Continuing in this way, $s_n \le \nu(1+\mu+\ldots+\mu^{n-1}) = \nu(\mu^n - 1)/(\mu - 1) = 1/4\epsilon c[(1+\epsilon d)^n - 1] \le 1/4\epsilon c[c^{\epsilon nd} - 1]$, where the second and fourth steps use facts from elementary algebra, and the third is substitution for μ and ν.

Define the efficiency of a partition $r_0, \ldots, r_n$ as above to be the number $\underline{r}/\epsilon n$ (i.e., a measure of how "nearly equally spaced the intervals are", clearly always between 0 and 1). Restricting ourselves to partitions with efficiencies at least 1/2, the formula of the previous paragraph becomes $s_n \le 1/4\epsilon c[e^{2\underline{r}d} - 1]$. Now consider a sequence of such partitions, each having with efficiency at least 1/2, each having an "n" twice that of its predecessor, and each being a refinement of its predecessor in the sense of the previous paragraph. Then, by this last formula, the sequence of endpoints obtained for each by our construction will be Cauchy (since this formula requires that the distance between two successive endpoints is bounded by some constant times the "ϵ" of the first partition), and hence will converge to some point of O. Repeating for different values r of $\underline{r}$ we obtain a mapping γ which assigns, to each sufficiently small positive r, a point $\gamma(r)$ of O.

We next claim that this mapping γ is C^1, and satisfies $D\gamma = \hat{\xi}\cdot\gamma$. Clearly, it suffices to check differentiability at O. We have $|\gamma(\underline{r}) - p - \underline{r}\hat{\xi}(p)| \le \underline{r}c(e^{2\underline{r}d} - 1)$, since the left side asks for the distance between the endpoints for a very fine partition, and the partition with just one step (i.e., using repeatedly the formula of the previous paragraph). But, as $\underline{r}$ goes to zero, $(e^{2\underline{r}d} - 1)$ also goes to zero, whence the left side is tangent at O. In other words, γ is differentiable, and $D\gamma = \hat{\xi}\cdot\gamma$. Since in particular γ is continuous, the right side of this last equation is continuous, whence $D\gamma$ is continuous, whence γ is C^1. Since γ is C^1, the right side is C^1, whence $D\gamma$ is C^1, whence γ is C^2. We continue in this way until we reach the differentiability class of $\hat{\xi}$. That is ... since γ is C^{p-1}, the right side is C^{p-1}, whence γ is C^p.

Now fix any point of our manifold M. Choosing a chart including this point, and using the construction above, we obtain an integral curve with this point as initial point. Extend this curve maximally by Zorn's Lemma. The resulting curve can have no endpoints (for if it had one, we could choose a chart including this endpoint, and use our construction to further extend the curve). This is our maximal integral curve.

This completes the proof of our theorem.

Conceptually, the proof above is extremely simple. For uniqueness, one uses the mean value theorem twice to show that "the two curves cannot get much farther apart than they had been previously". For existence, one constructs "broken straight lines, each segment of which is in the direction of the field at its initial point", and then takes a limit to obtain integral curves. I find it a bit strange that such simple ideas turn out to be so complicated when written out in detail. In the more conventional proof of this theorem, one rewrites "is an integral curve" as an integral equation, regards one side of that equation as a mapping from curves to curves, shows that the set of curves forms a complete metric space and that this mapping is a contraction mapping, and uses the contraction lemma. This proof is essentially the same as the one above. We have incorporated the definitions of "integral" and "contraction mapping" into the proof without mentioning these terms, and have proved directly in context the properties needed of integrals and contraction mappings. The above proof, while somewhat longer than that with integrals, would perhaps be comparable in length if the latter were accompanied by the definition and properties of integrals. The above proof appeals somewhat more to my intuition.

We first give a simple example just to illustrate how one uses the theorem.

Example. Two integral curves on a manifold cannot cross (i.e., have just one point in common), for, were this the case, one could reparametrize them so that the "crossing point" is the initial point of each, witch would violate uniqueness.

In the finite-dimensional case, the theorem above might be called "the fundamental existence and uniqueness theorem for solutions of systems of ordinary differential equations". We give one example to illustrate the appropriateness of this description, leaving the statement of the general case as an (easy) exercise.

Example. Consider the following system of ordinary differential equations. We are interested in functions x, y and z of real variable t, satisfying the equations

$$\ddot{x} = (1 - \dot{z}^2 + y^2)^{-1} + t^2$$
$$\dot{y} = -\dot{x}^2 + x^2 + \log(z + \dot{x}$$
$$\ddot{z} = \cos(2x\dot{z}t)$$

where "dot" means "d/dt". [Note that this is a "reasonable system", in the sense that the equations are solved for the highest derivative of each independent variable.] Denote by E the Banach space R^6, so a point of E is represented by six real numbers, $(x_1, \ldots, x_6)$. [Motivation: Think of $x_1 = x, x_2 = \dot{x}, x_3 = y, x_4 = z, x_5 = \dot{z}, x_6 = t$.] Next, let M be the (open sub-)manifold (of E) consisting of $(x_1, \ldots, x_6)$ satisfying $(1-(x_5)^2+(x_3)^2) \neq 0$ and $(x_4+x_2) > 0$. [Motivation: We throw away precisely those points at which the right sides of the differential equations above are not well-behaved.] Next, let ξ be the mapping from M to E with action $\xi(x_1, \ldots, x_2) = x_2, (1 - (x_5)^2 + (x_3)^2)^{-1} + (x_6)^2, -(x_2)^2 + (x_1)^2 + (x_3)^2 + \log(x_4 + x_2), x_5, \cos(2x_1x_5x_6), 1)$. [Motivation: That the second, third, and fifth entries are what they are can be seen by looking at the right sides of our differential equation. For the remaining three entries, append the equations "$\dot{x} = \dot{x}$", "$\dot{z} = \dot{z}$", and "$\dot{t} = 1$" to our three differential equations.] This ξ is (with respect to the natural chart on M, the component of) a C^∞ vector field on M.

A curve on M maps some open interval (a, b) of the reals to M. Hence, for γ such a curve and r in (a,b), $\gamma(r)$ is a point of M, a point which can be represented by a six-tuple. That is to say, a curve can be described by six functions, $x_1, \ldots, x_6$, of one real variable r. The statement that this γ be an integral curve of our vector field is, of course, the statement that $dx_1/dr = x_2, dx_2/dr = (1-(x_5)^2+(x_3)^2)^{-1}+(x_6)^2, dx_3/dr = -(x_2)^2+(x_1)^2+(x_3)^2+\log(x_4+x_2), dx_4/dr = x_5, dx_5/dr = \cos(2x_1x_5x_6), dx_6/dr = 1$. We now claim that every such integral curve gives rise to a solution of our differential equation (namely, given an integral curve, replace r by t, and identify $x(t)$ with $x_1(r)$, $y(t)$, with $x_3(r)$, $z(t)$ with $x_4(r)$, and t with $x_6(r)$. Then the first, fourth, and sixth of the equations above identify $x_2(r)$ with $\dot{x}(t)$, $x_5(r)$ with $\dot{z}(t)$, and $x_6(r)$ with t. The other three equations are then precisely our original differential equations.) Conversely, any solution of our original system of differential equations gives rise to an integral curve, by the same identifications. Under our identification, specifying an initial value for an integral curve amounts to specifying initial values for x, $\dot{x}$, y, z, $\dot{z}$ and t. From our theorem, we conclude therefore: Given initial values of x, $\dot{x}$, y, z, $\dot{z}$ and t, there is one and only one solution of the system of differential equations on the previous page, maximally extended.

I regard it as an enormous conceptual simplification that "everything one always wanted to know about ordinary differential equations" is summarized so concisely by the theorem. One might expect to be able to do a similar thing for (e.g., hyperbolic) partial differential equations, where now the manifold M is the infinite-dimensional manifold of possible initial data for the equation. I feel that it would be of particular interest to investigate (e.g., beginning with some examples) such a program.

21. Geometry of Lie Derivatives

We now provide the promised geometrical interpretation of Lie derivatives. It turns out that the material of this section is normally used only to obtain a quick general idea of what the Lie derivative of something is or is like, rather than being used directly in proofs. For this reason, the intuitive idea is more important than the details. In order to save time, we shall take advantage of this circumstance as follows: Although the claims of this section will be both precise and true, we shall merely sketch the proofs.

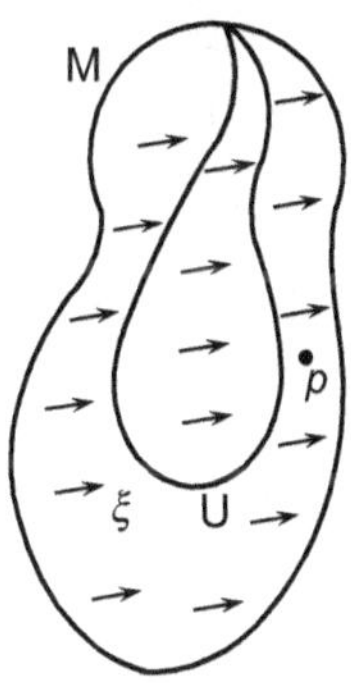

Fix, once and for all, a C^p $(p \geq 2)$ manifold M based on Banach space E, together with C^{p-1} E–field ξ on M. Fix also positive number a and open subset U of M such that the following property is satisfied: for each point p of U, the maximal integral curve of ξ with initial value p is defined at least on $(-a, a)$. Thus, in the example of the figure, the "U" shown would do the job. However, we could not in this example choose $U = M$, for the maximal integral curve with initial point p shown is not defined for r–values ip to a. Denote by I the open interval $(-a, a)$. Next, denote by φ the following mapping from $I \times U$ to M: For (r, p) in $I \times U$ so r is in $(-a, a)$ and p is in U), $\varphi(r, p)$, is the point $\gamma(r)$ of M, where γ is the maximal integral curve of ξ with initial point p. [The condition on U and a above is now seen as having been necessary in order that this φ be well-defined.] Thus, for example, we have $\varphi(0, p) = p$ for every p in U, and $\varphi(r, \varphi(r', p)) = \varphi(r + r', p)$ whenever both sides are defined.

We next note that, since both I and U are manifolds (as open subsets of manifolds), so is $I \times U$. Hence, φ is a mapping from one C^p manifold $(I \times U)$ to another (M). We claim: This φ is a C^p mapping of manifolds. To prove this, one first passes to a chart: Let O be an open subset of $E, \hat{\xi}$ a C^{p-1} mapping from O to E, U an open subset of O as above, and $\hat{\varphi}$ the mapping from $I \times U$ to O as above. We first show continuity. We have $|\hat{\varphi}(r', p') - \hat{\varphi}(r, p)| = |\hat{\varphi}(r', p') - \hat{\varphi}(r, p')| + |\hat{\varphi}(r, p') - \hat{\varphi}(r, p)|$. Set $c = \text{lub}\,|\hat{\xi}|$ and $d = \text{lub}\,|\mathrm{D}\hat{\xi}|$. The first term on the right above is $|\hat{\gamma}(r') - \hat{\gamma}(r)|$, where $\hat{\gamma}$

is the maximal integral curve with initial point p'. Hence, this first term is less than or equal to $c\,|r' - r|$. For the second term, set $s = |p' - p|$. Partition the open interval $(0, r)$ into n segments of maximum length ϵ as on page 99. Let $s_1, s_2, \ldots, s_n$ and μ and ν be as on that page. Then we have $s_1 \leq \mu s + \nu$, whence $s_n \leq \mu s_1 + \nu \leq \mu^2 s + \nu(\mu + 1)$, etc., to $s_n \leq \mu^n s + \nu(\mu^{n-1} + \ldots + 1)$. That is to say, $s_n \leq \epsilon^{\epsilon nd} s + \epsilon c/4(e^{\epsilon nd} - 1)$. Letting n go to infinity, keeping the efficiency greater than $1/2$, we obtain $|\hat{\varphi}(r, p') - \hat{\varphi}(r, p)| \leq e^{2rd} s$. Thus, we have obtained $|\hat{\varphi}(r', p') - \hat{\varphi}(r, p)| \leq c\,|r' - r| + e^{2rd}\,|p' - p|$, from which continuity follows immediately. We next show differentiability of $\hat{\varphi}$ at $(0, p)$. We have $|\hat{\varphi}(r', p') - \hat{\varphi}(0, p) - (p' - p) - r'\hat{\xi}(p)| = |\hat{\varphi}(r', p') - p' - r'\hat{\xi}(p)| \leq |\hat{\varphi}(r', p') - p' - r'\hat{\xi}(p')| + r'\,|\hat{\xi}(p') - \hat{\xi}(p)| \leq |\hat{\varphi}(r', p') - p' - r'\hat{\xi}(p')| + r'd\,|p' - p|$, where we used $\hat{\varphi}(0, p) = p$ in the first step, and the mean value theorem in the third. But we have already shown that the first term on the right is tangent at $r' = 0, p' = p$ while the second is obviously tangent. Hence, $\hat{\varphi}$ is differentiable at $(0, p)$, and $D\hat{\varphi}(0, p)$ is the mapping from $\mathbb{R} \times E$ to E which takes (b, x) to $x + b\hat{\xi}(p)$. Along similar lines, one shows that $\hat{\varphi}$ is differentiable everywhere, and then that it is C^p.

The result above, smoothness of $\hat{\varphi}$, is the basic result on the dependence of solutions of differential equations on the initial point and on the parameter. It states that "where you are in M after going amount r along the integral curve with initial point p depends smoothly on both r and p". Indeed, without such a result it is unlikely that differential equations would be of much use in physics. Think of M as representing the "space of states of some physical system", and of the parameter in the integral curves as representing "time". Then the integral curves represent "the evolution of the system through a succession of physical states with time". The initial point is "the state of the system at time zero". Now, one cannot avoid some small error in assigning to one's actual physical system an initial point (for, e.g., meters can only be read to a certain accuracy). If one is to do physics sensibly, it had better be true that these small errors in the assignment of initial point result only in small errors in one's prediction of what the system will be like at later times. That is to say, $\hat{\varphi}$ had better be continuous.

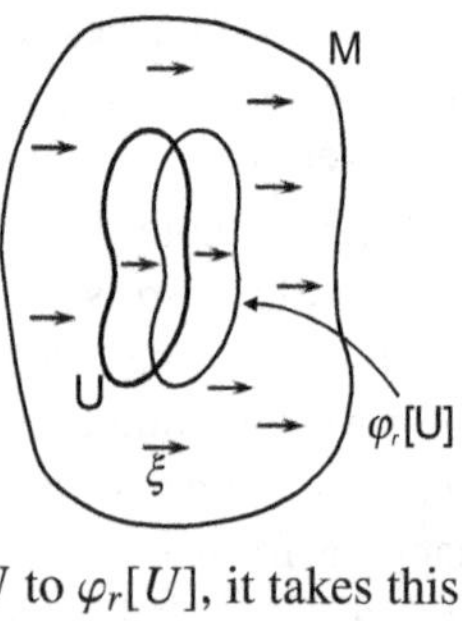

Let M, p, E, ξ, U, a, and φ be as above. Let, for each r in $(-a, a)$, φ_r be the mapping from U to M with action $\varphi_r(p) = \varphi(r, p)$. This φ_r, then, is the mapping which "slides each point of U an amount r along its integral curve". Each φ_r is a diffeomorphism from U to $\varphi[U]$ (for $\varphi_r \cdot \varphi_{-r}$ and $\varphi_{-r} \cdot \varphi_r$ are the identity wherever defined). Now let, in addition, α be some C^{p-1} A–field on M. Restricting α to the open submanifold U of M, we obtain an A–field $\underset{\sim}{\alpha}$ on U. Since φ_r is a diffeomorphism from U to $\varphi_r[U]$, it takes this $\underset{\sim}{\alpha}$ to some A–field on $\varphi - r[U]$. Next, fix a point p of U. Then for all sufficiently small r, p will be in $\varphi_r[U]$. In particular, the A–field we have defined on $\varphi_r[U]$ will determine an A–tensor at p. That is to say, we have obtained for each sufficiently small r, an A–tensor at p, which we denote $\alpha'(r)$. We can regard this α', then, as a curve in the Banachable space of A–tensors at p. [Intuitively, we "drag our original A–field α along by ξ, continually evaluating at p during the dragging, thus obtaining a one-parameter family of A–tensors at p".] Note, e.g., that $\alpha'(0) = \alpha(p)$ (since φ_0 is the identity), i.e., our curve has initial point which is just the value of the original A–field at p. Since φ is C^p, this curve is also C^p.

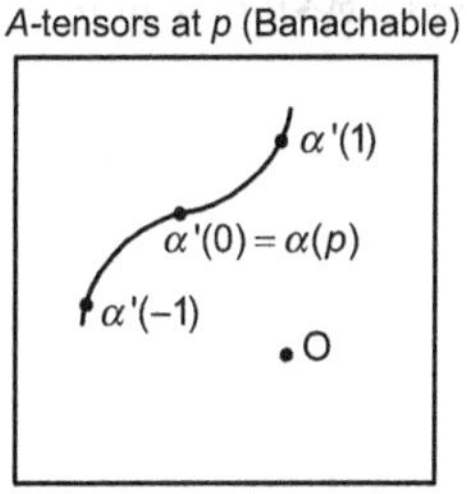

We now claim: The tangent vector to this curve α' at 0 is precisely $\mathcal{L}_\xi\alpha$, evaluated at p. That is to say, $\mathcal{L}_\xi\alpha$ at p is just "minus the rate of change of α as it is dragged along by ξ and evaluated at p". To prove this claim, one again passes to a chart. Let us suppose for a moment that one could manage to find a chart, containing p, in which the component $\hat{\xi}$ of ξ is constant, say ξ_0 (in E). Then, with respect to this chart, the component of $\mathcal{L}_\xi\alpha$ at p would be just $D\,\hat{\alpha}(\xi_0)$ (since other term in the definition of the Lie derivative involves $D\hat{\xi}$, which here vanishes: x is the point of E corresponding to p via our chart). In terms of this chart, however, our curve in the tensor space A is just $\alpha'(r) = \hat{\alpha}(x - r\xi_0)$ (for, in this chart, $\hat{\xi}$, is constant, whence its integral curves are straight lines). Thus, we have in this case that the component of the tangent to the curve α' is precisely the component of $\mathcal{L}_\xi\alpha$ at p, i.e., we have that our claim is true.

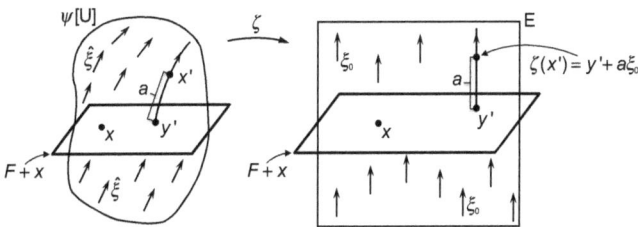

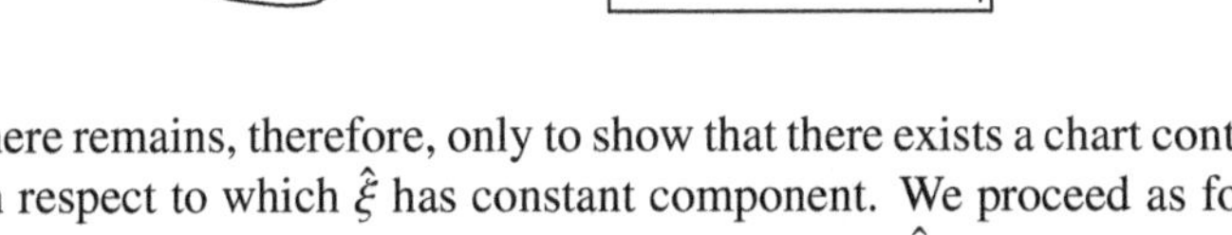

There remains, therefore, only to show that there exists a chart containing p with respect to which $\hat{\xi}$ has constant component. We proceed as follows. First, choose any chart, U, ψ, containing p, and let $\hat{\xi}$ be the component of ξ. Next, choose, a subspace F of E complementary to the vector $\xi_0 = \hat{\xi}(x)$. We introduce a mapping ζ from $\psi[U]$ to E as follows: For x' in ψU set $\zeta(x') = y' + a\xi_0$, where y' is the point of $\psi[U]$ at which the integral curve through x' meets $x + F$, and a is the parameter-difference along this curve from y' to x'. [By choosing U sufficiently small, this mapping will be well-defined.] Thus, for example, $\zeta(x) = x$. This is a C^p mapping with C^p inverse (by smoothness of φ and the inverse function theorem). Now set $\psi' = \zeta \cdot \psi$. Then we obtain a chart on M with respect to which the component of ξ is constant (since, in terms of this new chart, the integral curves of ξ are "straight lines", by construction).

We conclude, therefore, that $\mathcal{L}_\xi \alpha$ can be interpreted as the operation of "taking the rate of change of the field at each point under the 'sliding along' induced by motion along ξ". It is, for example, obvious from this interpretation that the Lie derivative of a natural tensor is zero, since natural tensors are invariant under diffeomorphisms. The other properties of Lie derivatives can be seen in a similar way.

22. Exterior Derivatives

We now consider the second "derivative-like notion" on a manifold. It turns out that the exterior derivative is applicable to only certain types of fields. We begin, therefore, with the study of these fields and their algebra.

Let E be any Banach space. Let n be any non-negative integer, and let σ be the following natural element of the tensor space $\mathcal{L}(\mathcal{L}(E,\ldots,E;\mathbb{R});\mathcal{L}(E,\ldots,E;\mathbb{R}))$ (with n E's on each side): For κ in $\mathcal{L}(E,\ldots,E;\mathbb{R})$, $\sigma(\kappa)$ is that element of $\mathcal{L}(E,\ldots,E;\mathbb{R})$ whose action on $x_1,\ldots,x_n$ in E is $\sigma(\kappa)(x_1,\ldots,x_n) = (1/n!)\Sigma(-1)^s\kappa(x_i,\ldots,x_j)$, where the sum extends over all $n!$ permutations of the vectors $x_1,\ldots,x_n$, where $x_i,\ldots,x_j$ is that permutation, and where s has the value +1 if the permutation is even and -1 if that permutation is odd. [For $n = 0, \sigma$ is the identity.]

Example. Set $n = 2$. Then $n! = 2$, and there are just two permutations of x_1, x_2, namely that which sends this to x_1, x_2 (even), and that to x_2, x_1, (odd). Hence, for κ in $\mathcal{L}(E,E;R)$, $\sigma(\kappa)$ has action $\sigma(\kappa)(x_1,x_2) = (1/2)(\kappa(x_1,x_2) - \kappa(x_2,x_1))$.

Example. Set $n = 3$. Then $\sigma(\kappa)(x_1,x_2,x_3) = (1/6)(\kappa(x_1,x_2,x_3)+\kappa(x_2,x_3,x_1)+\kappa(x_3,x_1,x_2)-\kappa(x_2,x_1,x_3)-\kappa(x_3,x_2,x_1)-\kappa(x_1,x_3,x_2))$. This σ will be called the *antisymmetrizer*, and its action that of *taking the antisymmetric part*.

We next claim that this σ has the following property: $\sigma\cdot\sigma = \sigma$, i.e., the antisymmetric part of antisymmetric part is the antisymmetric part. To prove this, let κ be in $\mathcal{L}(E,\ldots,E;\mathbb{R})$. Then $\sigma(\sigma(\kappa))(x_1,\ldots,x_n) = (1/n!)\Sigma(-1)^s \sigma(\kappa)(x_i,\ldots,x_j) = (1/n!)\Sigma(-1)^s[(1/n!)\Sigma(-1)^{s'}\kappa(x_{i'},\ldots,x_{j'})]$, in an obvious notation, where we have used the definition twice. The double sum on the right consists of $(n!)^2$ terms, each of which is $(1/n!)^2$ times plus or minus κ applied to some permutation of $x_1,\ldots,x_n$, plus if that permutation is even, minus if odd. But there are only $n!$ permutations of $x_1,\ldots,x_n$. Thus, each permutation occurs $(n!)$ times in this sum. Combining the terms corresponding to the same permutation, we obtain $(1/n!)^2\,\Sigma(-1)^s\kappa(x_i,\ldots,x_j)(n!)$, where the sum is over permutations, and where the last $n!$ arises because $n!$ terms of the original sum yield a single term of this one. But this last is precisely $\sigma(\kappa)(x_1,\ldots,x_n$. Thus, we have $\sigma(\sigma(\kappa))(x_1,\ldots,x_n) = \sigma(\kappa)(x_1,\ldots,x_n)$, whence since the x's and κ are arbitrary, we have $\sigma\cdot\sigma = \sigma$. [This argument is much easier than it looks. One should try it explicitly, e.g., for $n = 2$.]

An element of $\mathcal{L}(E, \ldots, E; \mathbb{R})$, ($nE$'s), κ is called an *n–form* if $\sigma(\kappa) = \kappa$. [A *0–form* is a number.] For example, every element of $\mathcal{L}(E; \mathbb{R})$ is a 1–form. From the result of the previous paragraph we have: for any κ in $\mathcal{L}(E, \ldots, E; \mathbb{R})$, $\sigma(\kappa)$ is an n–form (for $\sigma(\sigma(\kappa)) = \sigma(\kappa)$). We also have the following property of n–forms: For κ an n–form, $x - 1, \ldots, x_n$ in E, and $x_i, \ldots, x_j$ a permutation of these, $\kappa(x_1, \ldots, x_n) = (-1)^s \kappa(x_i, \ldots, x_j)$, where s is +1 if $x_i, \ldots, x_j$ is an even permutation of $x_1, \ldots, x_n$, and -1 if odd. [Proof: This is immediate from the definition of σ, for precisely the same terms appear in $\sigma(\kappa)(x_1, \ldots, x_n)$ as in $\sigma(\kappa)(x_i, \ldots, x_j)$, although all the terms of the latter will differ in sign from those of the former if $s = -1$.]

Our concern, for the moment, is the algebra of these n–form (a subject usually called exterior algebra). We next introduce a certain product between these forms. Fix non-negative integers n and $n!$ and let τ be the following natural element of $\mathcal{L}(\mathcal{L}(E, \ldots, E; \mathbb{R}), \mathcal{L}(E, \ldots, E; \mathbb{R}); \mathcal{L}(E, \ldots, E\mathbb{R}))$ (with n, n' and $n+n'$ E's, respectively): For κ in $\mathcal{L}(E, \ldots, E; \mathbb{R})$ (n E's) and κ' in $\mathcal{L}(E, \ldots, E; \mathbb{R})$ (n' E's), let $\tau(\kappa, \kappa)$ be that element of $\mathcal{L}(E, \ldots, E; \mathbb{R})$ ($n+n'$ E's) with $\tau(\kappa, \kappa')(x_1, \ldots, x_{n+n'}) = \kappa(x_1, \ldots, x_n)\kappa'(x_{n+1}, \ldots, x_{n+n'})$. Thus, τ simply "takes the product of the multilinear mappings" in the natural way. Of course, if κ and κ' happen to be n– and n'–forms, respectively, it will not in general be true that $\tau(\kappa, \kappa')$ is an $n + n')$–form. [For example, let κ and κ' be nonzero 1–forms. Then $\tau(\kappa, \kappa')(x_1, x_2) = \kappa(x_1)\kappa'(x_2)$, which will not be equal to -$\tau(\kappa, \kappa')(x_2, x_1)$, whence $\tau(\kappa, \kappa')$ will not be a 2–form.] We can, however, obtain an $(n + n')$–form by applying σ. Thus: for κ and κ' n– and n'–forms, respectively, the $(n + n')$–form $\sigma(\tau(\kappa, \kappa'))$ is written $\kappa \wedge \kappa'$ and is called the *wedge product* of κ and κ'. [For either n or n' zero, one multiples the form by the number.]

We have now defined forms, together with a product operation on them. We wish next to obtain properties, of which there are three. First, we have that the wedge product is linear in each factor, i.e., $(\kappa + \underline{\kappa}) \wedge \kappa' = \kappa \wedge \kappa' + \underline{\kappa} \wedge \kappa'$ and $\kappa \wedge (\kappa' + \underline{\kappa'}) = \kappa \wedge \kappa' + \kappa \wedge \underline{\kappa'}$, where κ and $\underline{\kappa}$ are n–forms and κ' and $\underline{\kappa'}$ are n'–forms. These facts, together with the fact that the sum of two n–forms is an n–form, are all immediate, since everything in sight linear. Second, we have; For κ, κ', and κ'', n–, n'–, and n''–forms, respectively, $(\kappa \wedge \kappa') \wedge \kappa'' = \kappa \wedge (\kappa' \wedge \kappa'')$ (each side an $(n + n' + n'')$–form). [Proof: First note that, for any κ and κ', $\sigma \cdot \tau(\sigma(\kappa), \kappa') = \sigma \cdot \tau(\kappa, \kappa')$, for, applying both sides to $(x_1, \ldots, x_{n+n})$ the innermost "σ" on the left will antisymmetrize over the first n "x's" in each term, which is unnecessary, since the outermost "σ" already antisymmetrizes over all "x's". Next, note that $\tau(\tau(\kappa, \kappa'), \kappa'') = \tau(\kappa, \tau(\kappa', \kappa''))$, which is immediate from the definition (applying to $x_1, \ldots, x_{n+n'+n''}$, κ will get the first n x's, κ' the next n', and κ'' the last n'', using either side of the equation). We now have $(\kappa \wedge \kappa') \wedge \kappa'' = \sigma \cdot \tau(\sigma \cdot \tau(\kappa, \kappa'), \kappa'')) = \sigma \cdot \tau(\tau(\kappa, \kappa', \kappa'') = \sigma \cdot \tau(\kappa, \tau(\kappa', \kappa'')) = \sigma \cdot \tau(\kappa, \sigma \cdot \tau(\kappa', \kappa'')) = \kappa \wedge (\kappa' \wedge \kappa'')$, where we used definitions in the first and fifth steps, our first observation above in the second and fourth, and our our second observation in the third.] Thus, the wedge product is associative. Finally, we claim a sort of commutativity: For κ and κ' $n-$ and $n'-$forms, respectively, $\kappa \wedge \kappa' = (-1)^{nn'}\kappa' \wedge \kappa$. [Proof: We have $\tau(\kappa, \kappa')(x_1, \dots, x_{n+n'}) = \kappa(x_1, \dots, x_n)\kappa'(x_{n+1}, \dots, x_{n+n'})$, and $\tau(\kappa', \kappa)(x_1, \dots, x_{n+n'}) = \kappa'(x_1, \dots, x'_n)\kappa(x_{n'+1}, \dots, x_{n'+n}) = \kappa(x_{n'+1}, \dots, n_{n'+n'})$ $\kappa'(x_1, \dots, x'_n)$. The order of the x's in the expression on the right in the second formula is an odd permutation of that in the first formula if and only if both n and n' are odd. Hence, $\sigma(\tau(\kappa, \kappa')) = (-1)^{nn'}\sigma(\tau(\kappa', \kappa))$.]

This, then, is the set-up: We have $n-$forms, and on them a bilinear, associative, and "more or less commutative" product. This algebra in tensor spaces goes over immediately to fields on a manifold M. Thus, an *$n-$form* (field) on M is an $\mathcal{L}(E, \dots, E; \mathbb{R})-$field whose component with respect to any chart at any point is an $n-$form. Since the wedge product is a natural operation, one can take the wedge product of $n-$forms on M. Since our three properties hold pointwise, they hold for the fields.

Let M be a C^p ($p \geq 2$) manifold based on Banach space E. The exterior derivative, which we now define, will be applicable only to (C^{p-1}) forms on M. Let, then, κ be a C^{p-1} $n-$form on M. Passing to a chart, we have an open subset O of E, and a C^{p-1} mapping $\hat{\kappa}$ from O to the tensor space $\mathcal{L}(E, \dots, E; \mathbb{R})$, where, for each x, $\sigma(\hat{\kappa}(x)) = \hat{\kappa}(x)$ (i.e., where each $\hat{\kappa}(x)$ is an $n-$form in our tensor space). Now fix x in O, and consider $D\hat{\kappa}(x)$, an element of $\mathcal{L}(E; \mathcal{L}(E, \dots, E; \mathbb{R}))$.

We next note that there is a natural isomorphism from $\mathcal{L}(E; \mathcal{L}(E, \dots, E; \mathbb{R}))$ (n E's on the right) to $\mathcal{L}(E, \dots, E; \mathbb{R})$ ($n+1$ E's), namely, that which sends τ in the former to that element of the latter which acting on $x_1, \dots, x_{n+1}$ in E, produces the real number $\tau(x_1)(x_2, \dots, x_{n+1})$. In what follows, action of this isomorphism will be assumed implicitly when appropriate. Thus, in particular, we may regard $D\hat{\kappa}(x)$ as an element of $\mathcal{L}(E, \dots, E; \mathbb{R})$ ($n+1$ E's). Hence, applying the antisymmetrizer, $\sigma(D\hat{\kappa}(x)$ is an $n+1)-$form. We now claim that, in fact, this construction yields a tensor at the point p of M. That is, let U, ψ and U', ψ' be two charts, with p in $U \cap U'$. Then, denoting by $\hat{\kappa}$ and $\hat{\kappa}'$ the respective components of κ, we have that, for $x_1, \dots, x_n$ in E, $\hat{\kappa}'(x')(\iota(x_1 \dots, \iota(x_n)) = \hat{\kappa}(x)(x_1, \dots, x_n)$, where $x = \psi(p)$, $x' = \psi'(p)$, and where $\iota = D(\psi' \cdot \psi^{-1})(x)$. Taking the derivative of each side of this equation (with respect to x), and applying to an arbitrary vector y in E, we have $D\hat{\kappa}'(x')(\iota(y))(\iota(x_1), \dots, \iota(x_n)) + \hat{\kappa}'(x')(D\iota(y)(x_1), \dots, \iota(x_n)) + \dots +$ $\hat{\kappa}'(x')(\iota(x_1), \dots, D\iota(y)(x_n)) = D\hat{\kappa}(x)(y)(x_1, \dots, x_n)$. Now apply, to each side of this equation, the antisymmetrizer (over the $(n+1)$ vectors $y, x_1, \dots, x_n$). Then all the terms except the first on the left vanish (for, e.g., the second term, we have $D\iota(y)(x_1) = D\iota(x_1)(y)$, by symmetry of mixed partials, while, in the sum resulting from application of the antisymmetrier, for each containing $D\iota(y)(x_1)$ there will be another term identical except for sign but with this replaced by $D\iota(x_1)(y)$. Hence, the terms will all cancel in pairs.) Thus, we obtain $\sigma(D\hat{\kappa}'(x'))(\iota(y),\iota(x_1),\ldots,\iota(x_n)) = \sigma(D\hat{\kappa}(x))(y,x_1,\ldots,x_n)$. But this is precisely the statement that $(\sigma(D\hat{\kappa}'(x'));U',\psi') \approx (\sigma(D\hat{\kappa}(x));U,\psi)$. Thus, we obtain an $(n+1)$–form at p.

Repeating the above at each point of M, we conclude: Given a C^{p-1} n–form (failed) on M, we obtain as above a C^{p-2} $(n+1)$–form (field on M). This latter is called the *exterior derivative* of κ, written $d\kappa$.

Example. Let $n = 0$. The 0–form κ is just a scalar field. The instructions above read in this case: The exterior derivative d is that 1–form (i.e., $\mathcal{L}(E;\mathbb{R})$–field such that, passing to a chart and letting y in E be arbitrary, $\widehat{d\kappa}(x)(y) = D\hat{\kappa}(x)(y)$ (antisymmetrization being unnecessary for 1–forms). But the right side of the above is precisely the directional derivative of κ in the y–direction. That is to say, $d\kappa$ is what one in elementary calculus calls the gradient. We conclude: On 0–forms, the exterior derivative is the gradient.

Example. Let $n = 1$. Then, for κ a 1–form on M, the exterior derivative has the following expression in terms of a chart: $(d\hat{\kappa}(x))(y,x_1) = D\hat{\kappa}(x)(y)(x_1) - D\hat{\kappa}(x)(x_1)(y))$. That is to say, one takes "half the derivative of the x_1–component of $\hat{\kappa}$ in the y–direction minus half the derivative of the y–component of $\hat{\kappa}$ in the x_1–direction". These instructions will be recognized as those which, in elementary calculus, yield the curl. On the 1–forms, therefore, the exterior derivative is the curl.

On higher forms, the exterior derivative is a sort of "generalized curl".

As usual, we now want to determine the various properties of the exterior derivatives. There are four. First, we have: For κ and λ n–forms, $d(\kappa+\lambda) + d\kappa + d\lambda$. Second: For κ and κ' n– and n'–forms, respectively, $d(\kappa\wedge\kappa') = d\kappa\wedge\kappa' + (-1)^n\kappa\wedge d\kappa'$. This is a sort of "Leibniz rule". Note that this equation makes sense, each side being an $(n+n'+1)$–form. [Proof: In terms of a chart, the left side, applied to $y,x_1,\ldots,x_{n+n'}$, is $\sigma(D(\hat{\kappa}(x_1,\ldots,x_n)\,\hat{\kappa}'(x_{n+1},\ldots, x_{n+n'}))(y)) = \sigma(D\hat{\kappa}(y)(x_1,\ldots,x_n)\,\hat{\kappa}'(x_{n+1},\ldots,x_{n+n'})) + \sigma(\hat{\kappa}(x_1\ldots,x_n)\, D\hat{\kappa}'(y)(x_n,\ldots,x_{n+n'}))$. But the two terms on the right in this formula are the chart-representations of the two terms on the right in our claimed equation.]

Example. Let $n = 0$, $n' = 1$. Then κ is a scalar field, κ' a 1–form, and our property above becomes $d(\kappa\kappa') = d\kappa\wedge\kappa' + \kappa d\kappa'$. This will be recognized as a formula from elementary vector calculus: The curl of a function times a vector is given by the gradient of the function cross the vector plus the function times the curl of the vector.

The third property is this: For any n–form κ, $dd\kappa = 0$). [Proof: In terms of a chart, the left side, applied to $z,y,x_1,\ldots,x_n$, is $\sigma D(D\hat{\kappa}(y))(z)(x_1,\ldots,x_n) = \sigma(DD\hat{\kappa}(y)(z)(x_1,\ldots,x_n))$, where the antisymmetrizer is over $z,y,x_1,\ldots,x_n$. Since mixed partial are symmetric, the antisymmetrizer annihilates.]

Example. Set $n = 0$ in the last property. Then, in elementary calculus terminology, we have: The curl of the gradient of a scalar field vanishes.

This, of course, is true.

The final property relates Lie and exterior derivatives. For this property, we need a bit of notation. For xi an E–tensor and κ an n–form, we write $\xi \cdot \kappa$ for the $(n-1)$–form whose action on $x_1, \ldots, x_{n-1}$ is $(\xi \cdot \kappa)(x_1, \ldots, x_{n_1}) = \kappa(\xi, x_1, \ldots, x_{n_1})$. Similarly for fields. [This, of course, is the action of a natural operation.] Thus, for example, we have $\xi \cdot (\xi \cdot \kappa) = 0$. The final property is this: For ξ an E–field and κ and n–form (field), $(n+1)\xi \cdot d\kappa - \mathcal{L}_\xi \kappa + n\, d(\xi \cdot \kappa) = 0$. [Note that this is well-defined, each term on the left being an n–form.] [Proof: Passing to a chart and applying to $x_1, \ldots, x_n$, $\xi \cdot d\kappa$ becomes $\sigma(D\hat{\kappa}(\hat{\xi})(x_1, \ldots, x_n))$ where the antisymmetrizer is over $\hat{xi}, x_1, \ldots, x_n$; $\mathcal{L}_\xi \kappa$ becomes $D\hat{\kappa}(\hat{\xi}(x_1, \ldots, x_n) + \hat{\kappa}(D\hat{\xi}(x_1, \ldots, x_n)) + \ldots + \hat{\kappa}(x_1, \ldots, D\hat{\xi}(x_n))$; and $d(\xi \cdot \kappa)$ becomes $\sigma(D(\hat{\kappa}(\hat{\xi}, x_2, \ldots, x_n))(x_1))$, where the antisymmetrizer is over $x_1, \ldots, x_n$. Expanding the antisymmetrizer in all three expressions, and the last expression using the Leibnitz rule, one sees that all the terms in the claimed combination cancel.]

Example. Setting $n = 0$, our last property becomes $\xi \cdot d\kappa - \mathcal{L}_\xi \kappa = 0$ for κ a 0–form. But this formula is true, for each of the two terms on the left has been seen to be the directional derivative of κ in the ξ–direction.

In the finite-dimensional case, the last property above provides a convenient way of defining the exterior derivative, without reference to charts. Rewrite the the formula of that property in the form $\xi \cdot d\kappa = 1/(n+1)\, \mathcal{L}_\xi \kappa - n/(n+1)\, d(\xi \cdot \kappa)$. Then, for $n = 0$, the right side is known (once one has Lie derivatives), and so this formula can be used to define the exterior derivative of 0–form. Once exterior derivatives have been defined on forms up to $(n-1)$, the right side of this formula, for κ an n–form, is known, whence this formula can be used to define the exterior derivative of an n–form. By induction, then, one obtains the definition of exterior derivative on any form. A similar program could be carried out in the infinite-dimensional case, but, as usual, additional complications arise from the feature that manifolds may not admit non-trivial fields globally.

23. Derivative Operators

We come, finally, to the third "type of derivative".

Fix a C^p ($p \geq 2$) manifold M based on Banach space E. Fix a point p of M. Consider now pairs, $(\Gamma; U, \psi)$, where U, ψ is a chart containing p, and Γ is an element of $\mathcal{L}(E, E; E)$ which is symmetric (i.e., which satisfies, for any y, y' in E, $\Gamma(y, y') = \Gamma(y', y)$). Given two such, we write $(\Gamma; U, \psi) \approx (\Gamma'; U', \psi')$ provided the following equation holds: For any vectors y and z in E, $\Gamma'(\iota(y), \iota(z)) = \iota(\Gamma(y, z)) + D\iota(x)(y, z)$, where, as usual, $\iota = D(\psi' \cdot \psi^{-1})$, and $x = \psi(p)$. Each side of this equation is an element of E. it should be noted that this equation is different from the corresponding equation for an $\mathcal{L}(E, E; E)$ tensor at p: Indeed, were the second term on the right above omitted, we would have precisely the tensor relation. The things we are now defining are "geometrical objects" which are somewhat like, but not precisely the same as, tensors.

We claim that "$\approx$" is an equivalence relation. Clearly, we have $(\Gamma; U, \psi) \approx (\Gamma; U, \psi)$. Suppose next that $(\Gamma; U, \psi) \approx (\Gamma'; U', \psi')$. Then we have $\Gamma'(\iota(y), \iota(z)) = \iota(\Gamma(y, z)) + D\iota(x)(y, z)$, from which it follows, we claim that $\Gamma(\iota^{-1}(y), \iota^{-1}(z)) = \iota^{-1}(\Gamma(y, z) + D(\iota^{-1})(x')(y, z)$, [Proof: Expand $D(\iota^{-1}$ in the second formula, and replace y by $\iota(y)$ and z by $\iota(z)$.] But this second formula is precisely the statement that $(\Gamma'; U', \psi') \approx (\Gamma; U, \psi)$. Finally, that $(\Gamma; U, \psi) \approx (\Gamma'; U', \psi')$ and $(\Gamma'; U', \psi') \approx (\Gamma''; U'', \psi'')$ implies $(\Gamma; U, \psi) \approx (\Gamma''; U'', \psi'')$ is, similarly, an easy exercise in algebra.

An equivalence class of such pairs is called a *connection* at p. A connection field on M is a mapping which associates with each point p of M a connected at p. The component of a connection field with respect to a chart is the corresponding mapping from $O = \psi[U]$, an open subset of E, to $\mathcal{L}(E, E; E)$. A connection field is said to be C^{p-2} if its component is C^{p-2} for every admissible chart. [Note that we here go from p all the way down to $(p-2)$. The reason is that "$D\iota$", the second derivative of the chart-mappings, appears in the formula for component-change under chart-change.] A C^{p-2} connection field on M is normally called a *derivative operator* on M.

The interest in the notion of a derivative operator stems from the feature that such an operator permits one to take the derivative of an arbitrary tensor field on M, again obtaining a tensor field. We next see how this comes about. Let α be a C^{p-1} A–field on M, and fix a derivative operator on M. Choose a chart, and let $\hat{\alpha}$ and $\hat{\Gamma}$ be the components of α and the derivative operator, respectively. Now let x be any point of $O = \psi[U]$, and y any vector in E, and consider the free A–tensor $D\hat{\alpha}(x)(y) - \kappa(\widehat{\Gamma}(x)(y), \hat{\alpha}(x))$, where κ is the element of $\mathcal{L}(\mathcal{L}(E;E),A;A)$ defined at the top of page 92. For each y in E, the above is a free A–tensor, clearly linear in y. Hence, the above can be regarded as a mapping from E to A, i.e., as a free $\mathcal{L}(E;A)$–tensor. Now change the chart. We have the formulae for how $\hat{\kappa}$ and $\hat{\Gamma}$ change, and so can write down how the free $\mathcal{L}(E;A)$–tensor above changes. We claim: It behaves like a $\mathcal{L}(E;A)$–tensor at p. The calculation is identical with that for Lie derivatives: One picks up a "$D\iota$" term from "$D\hat{\alpha}$", and also a "$D\iota$" term from $\widehat{\Gamma'}$ (by the formula for the component-change under chart-change for a connection). The expression above has been adjusted so that these terms cancel. Thus, we obtain an $\mathcal{L}(E;A)$–tensor at p, which we write $\nabla\alpha(p)$. Repeating for each point of M, we obtain a C^{p-2} $\mathcal{L}(E;A)$– field on M, $\nabla\alpha$. This field is called the *derivative* of α (with respect to our given derivative operator).

Example. Let α be a scalar field. Then $\nabla\alpha$ must be a $\mathcal{L}(E;\mathbb{R})$–field, i.e., a 1–form. In this case, the "κ–term" above vanishes, and we have $(\widehat{\nabla\alpha})(y) = D\hat{\alpha}(y)$. But this is precisely the formula for the exterior derivative. Hence, the result of application of a derivative operator to a scalar field is to take its exterior derivative.

Thus, the derivative, $\nabla\alpha$, of a C^{p-1} A–field α is a C^{p-2} $\mathcal{L}(E;A)$–field. This is of course what one might have expected: The "E" refers to the possible directions in M along which the derivative might be taken.

We obtain a few properties of derivatives via derivative operators. Clearly, $\nabla(\alpha + \alpha') = \nabla\alpha + \nabla\alpha'$. Next, let τ be a natural tensor field. Then, for its component with respect to a chart, $\hat{\tau}$ is a constant mapping. Hence, $D\hat{\tau} = 0$. Furthermore, as we saw in Sect. 19, $\kappa(\hat{\Gamma}, \hat{\tau}) = 0$ in this case. from the formula above for the derivative, therefore, we conclude that $\nabla\tau = 0$. The derivative of a natural field vanishes. We next note that, for ρ any $\mathcal{L}(A_1, \ldots, A_n; B)$–field, and $\alpha_1, \ldots, \alpha_n$ $A_1, \ldots, A_n$–fields, we have $\nabla(\rho(\alpha_1, \ldots, \alpha_n)) = (\nabla\rho)(\alpha_1, \ldots, \alpha_n) + \rho(\nabla\alpha_1, \ldots, \alpha_n) + \ldots + \rho(\alpha_1, \ldots, \nabla\alpha_n)$. it follows immediately that the derivative satisfies the Leibnitz rule for any natural operation on tensor fields. Finally, we show that "mixed derivatives commute", at least when applied to scalar fields. Let f be a scalar field. Then $\nabla\nabla f$ is a $\mathcal{L}(E,E;R)$–field. We claim that this $\nabla\nabla f$ is symmetric (i.e., $\nabla\nabla f(x_1, x_2) = \nabla\nabla f(x_2, x_1)$). Indeed, choosing a chart, we have $\widehat{\nabla f}(x_1) = D\hat{f}(x_1)$, whence $\widehat{\nabla\nabla f}(x_1, x_2) = DD\hat{f}(x_1, x_2) - Df(\widehat{\Gamma}(x_1, x_2))$. But the first term is symmetric in x_1 and x_2 since mixed partials commute, while the second is symmetric since $\widehat{\Gamma}$ is.

Again, in the finite-dimensional case one normally defines derivatives operators by their properties, rather than via charts, a method which seems to be more awkward in the infinite-dimensional case.

Derivative operators are a rather "brute force" way of making available a derivative. One simply introduces what one needs (a connection field) to be able to eliminate "$D\iota$–terms" when taking component-derivatives. As one might expect, then, derivative operators will normally not be very useful unless one arises naturally from what one has available to him already. It turns out that there is one situation in which such an operator does appear naturally: in the presence of a metric. Thus, it is normally when one deals with manifolds-with-metrics, i.e., with what is called Riemannian geometry, that derivative operators play the important role. In a similar way, Lie derivatives are not normally very useful unless one happens to have some natural vector field around.

Finally, we remark that there are numerous relationships between Lie derivatives, exterior derivatives, and derivatives via derivative operators. In particular, the former two can be expressed in terms of the latter.

24. Derivatives: Summary

The table below summarizes selected features of the three types of derivative we have discussed.

Name	Lie Derivative	Exterior Derivative	Derivative
Applies to	Any field α	Forms only	Any field α
Requires	E–field ξ	Nothing	Derivative operator ∇
Symbol	$\mathcal{L}_\xi \alpha$	$\mathrm{d}\alpha$	$\nabla\alpha$
Properties	Additive Leibnitz	Additive "Leibnitz"	Additive Leibnitz
On Natural Fields	Gives zero	—	Gives zero

25. Partial Differential Equations

A vector field on a manifold has unique maximal integral curves. As we have seen in Sect. 20, this fact leads, in the finite-dimensional case, to a simple, geometrical statement of the existence and uniqueness of solutions of systems of ordinary differential equations. But the fact is true also in the infinite-dimensional case. It is natural to ask, therefore, whether or not one can, in a similar way, obtain a "geometrization" of certain partial differential equations. It is our goal in this section to look briefly, by means of an example, at this question. Our conclusion, essentially, will be that naive ideas do not work.

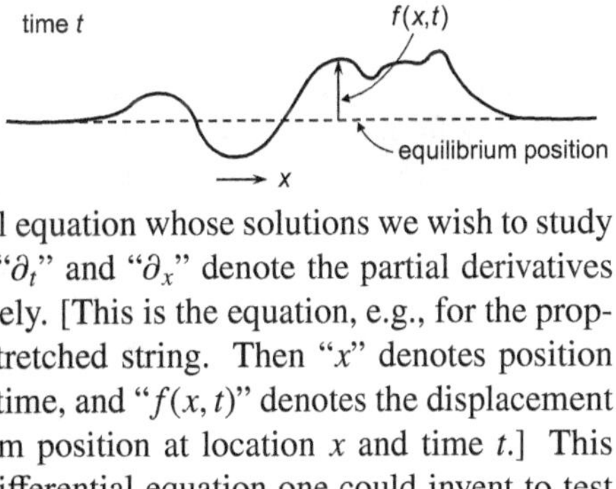

We shall be concerned with mappings from $\mathbb{R}^2$ to $\mathbb{R}$, i.e., with real-valued functions of two real variables. We denote such a function f, and the variables t and x. The differential equation whose solutions we wish to study is this: $(\partial_t)^2 f = (\partial_x)^2 f$, where "$\partial_t$" and "$\partial_x$" denote the partial derivatives with respect to t and x, respectively. [This is the equation, e.g., for the propagation of waves on a tightly stretched string. Then "x" denotes position along the string, "t" denotes the time, and "$f(x, t)$" denotes the displacement of the string from its equilibrium position at location x and time t.] This is perhaps the simplest partial differential equation one could invent to test our program: It is linear, hyperbolic, of second order, and in two variables. One of the nice features of this particular equation is that it is easy to write down its general solution. Indeed, let p and q each be functions of one real variable.

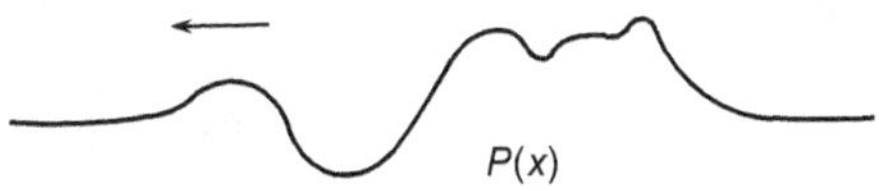

Then, clearly, the function f with action $f(x, t) = p(x + t)$ is a solution of our equation, as is the function f with action $f(x, t) = q(x - t)$. By linearity, their sum, $p(x+t)+q(x-t)$, is also a solution. [Physically, a solution of the form $p(x+t)$ corresponds to a wave pulse on the string, of shape described by p, which moves to the left along the string without changing its shape. Similarly, a solution of the form $q(x-t)$ corresponds to a pulse moving to the right.] It turns out that, in fact, the most general solution of our partial differential equation is of the form $p(x+t)+q(x-t)$. [A statement which is true, whose proof we omit, but which, as we shall see shortly, could very easily be proven after completion of the program we are here discussing.]

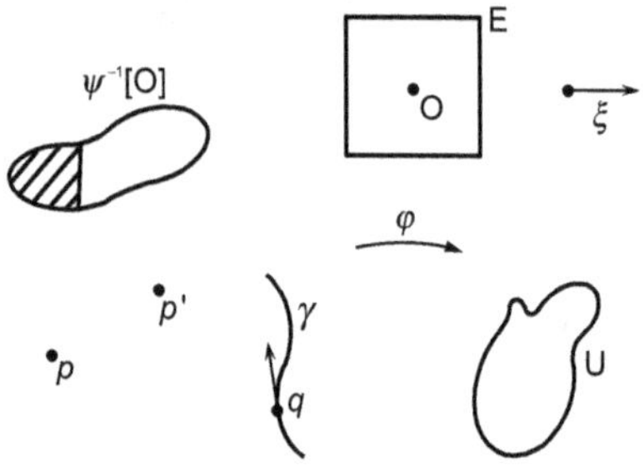

The first step in the geometrization of an ordinary differential equation was to cast it into the language of manifolds. We now wish to do the same thing for the present equation. To this end, let each of E_1 and E_2 be a "space" (eventually, it is hoped, a Banach space) of functions of one real variable, and let E denote their product, $E = E_1 \times E_2$. Consider now a solution f of our partial differential equation. For each real number t_0, let g_{t_0} be the function with action $gt_0(x,t) = f(x,t_0)$ and h_{t_0} the function with action $h_{t_0}(x) = (\partial_t f)(x,t_0)$. [Physically, g_{t_0} describes the "shape" of the string at time t_0, and h_{t_0} the "velocity of motion up or down of each segment of the string at time t_0".] The pair (g_{t_0}, h_{t_0}), then, as a pair of functions of one real variable, should define a point of $E_1 \times E_2$, i.e., a point of E. Since this is the case for each real number t_0, we obtain a mapping γ from the reals to E, with action $\gamma(t_0) = (g_{t_0}, h_{t_0})$. That is, we obtain a curve in E. Of course, given this curve in E, the original solution f of the partial differential equation can be recovered immediately, for, e.g., $f(x,t)$ is then the first entry of $\gamma(t)$, evaluated at x. In this way, then, we can describe solutions of our partial differential equation by certain curves. [Physically, E is the space of "possible physical states of the string", and the curve describes "the evolution of the string with time through a succession of physical states".]

The general idea is this. We wish to introduce a certain E–field on the (eventually, a manifold) E, so constructed that its integral curves are precisely the curves, obtained above, corresponding to solutions of our partial differential equation. There is a simple check one can do to see if this is even close to being possible: One can determine whether a single (e.g., initial) point on such a curve determines the curve (i.e., the solution of our equation) uniquely. That is, we ask: Is it true that, given any (g,h) in E, there is one and only one solution of our partial differential equation, f with $g(x) = f(x,0)$ and $h(x) = (\partial_t f)(x,0)$. In fact, the answer is yes, and the unique solution is given by $f(x,t) = 1/2(g(x+t)+g(x-t))+1/2(h(x+t)-h(x-t))$. [The right side is indeed a solution, as we have already observed. Furthermore, as one easily checks directly, the f defined by this equation indeed satisfies $g(x) = f(x,0)$

and $h(x) = (\partial_t f)(x, 0)$. That this is the unique f satisfying these conditions is also clear, since every solution f is of the form $p(x + t) + q(x - t)$.] Thus, our "solution curves" do not cross.

We now have our "space" E, and our curves, and we next wish to find a vector field whose integral curves are these curves. But this is quite easy: We simply take the tangent vectors to our curves. Thus, let f be a solution of our partial differential equation, γ the corresponding curve in E, and let $\gamma(0) = (g, h)$. Then γ has action $\gamma(t) = (g_t, h_t)$, with $g_t(x) = f(x, t)$ and $h_t(x) = (\partial_t f)(x, t)$, whence the tangent vector to γ at zero is $(\hat{g}, \hat{h})$, where $\hat{g}(x) = (\partial_t f)(x, 0)$, and $\hat{h}(x) = (\partial_t)^2 f(x, 0)$. In other words, $g = h$ and $h = g''$ (where prime denotes derivative with respect to the single variable, x on which g depends) this last equation comes from the partial differential equation on f). This little calculation of the tangent vector also tells us what the vector field should be. Let ξ be the mapping from E to E with action $\xi(g, h) = (h, g'')$. [Note what happened here: The tangent vector depended only on (g, h), so it could be represented by some vector field.] Then the integral curves of this γ will be precisely the solution curves of our partial differential equation.

The above discussion is essentially only motivation, for we have omitted the "detail" of specifying precisely which functions are permitted to be in the space E. If, however, this single point could be cleared up, we would have a geometrization of this particular (and then, presumably, also of many similar) partial differential equations. We now concern ourselves with this issue of specifying E.

A natural choice might be this. Choose integer $p \geq 0$, and let E_1 consist of C^p functions g with lub $|g|$ + lub $|g'|$ + ... + lub $|g^{(p)}|$ finite, where the norm is this sum. As we have seen, this E_1 is indeed a Banach space. Repeating for E_2, we obtain a Banach space E.

We first observe that, since ξ is to be a mapping from E to E, since ξ is to have action $\xi(g, h) = (h, g'')$, it must be true that, for any (g, h) in E, so is (h, g''). In particular, we must choose $p \geq 2$, in order that "g''" make sense. But even this is not enough, for, having so chosen p, each of h and g'' must be C^p – in general, the latter will be only C^{p-2}. One might hope to get around this problem by, e.g., letting E_1 consist of C^p functions and E_2 of C^{p-2}. Then g will be C^p, whence g'' be C^{p-2}, but this g'' is the second entry of (h, g''), and functions there need only be C^{p-2} (since they are in E_2). This, however, will not work either, for now the "h" in (g, h) will only have to be C^{p-2}, whence "h" is not a suitable first entry in (h, g''). We conclude, therefore, that there seems to be reasonable choices for E_1 and E_2 as spaces of functions of finite differentiability class for which our "ξ" will indeed be a well-defined mapping from E to E. The problem, of course, is that ξ "takes derivatives", which tends to "push us out of the space in which we began".

The obvious solution is to pass to C^∞ functions (for which "take a derivative" does not get one out of the space). Suppose, then, that we let E_1 consists of C^∞ functions whose value, the value of whose derivative, etc.s all bounded. We could not choose on this E_1 a norm of the form lub $|g|$ + lub $|g'|$ + ... + lub $|g^{(p)}|$, for some fixed positive p, for then E_1 would not be complete. [As a general rule, completeness fails when restrictions are imposed on the class of admitted functions which are not suitably incorporated into the norm.] We could, however, set $d(g) = \text{lub}\,|g|/(1 + \text{lub}\,|g|) + 1/2\text{lub}\,|g'|/(1 + \text{lub}\,|g'|) + 1/4\text{lub}\,|g''|/(1 + \text{lub}\,|g''|) + \ldots$, a measure of the "size" of g. [The idea of this expression is that each term, without its numerical factor, is less than one, whence the nth term is less than $1/2^n$, whence the sum always converges.] This is indeed a metric on our set of C^∞ functions, i.e., letting the distance between g and $\tilde{g}$ be $d(g - \tilde{g})$. Furthermore, as one might expect from the appearance of the formula above, our E_1 is complete under this metric. Our E_1 how includes the derivative of each function in it, and is complete. Unfortunately, we have also now introduced a new problem: The "d" above is not a norm, for it fails to be true that $d(ag) = a\,d(g)$ for a real number. [This E_1 is what is called a Frechet space, about which we shall say slightly more later.] The problem, of course, is all those "1 + lub$''$'s in the denominators.

To summarize, we have found so far that C^p functions will not work for E, because this set of functions does not include the derivative of each of its elements. Furthermore, C functions with "C^p–norm" will not work either (although derivatives are not included) because completeness fails. "Completeness" can be restored for C^∞ functions, but the result is just a metric rather than a norm.

Let us try, then, to see if we can find some norm on C^∞ functions. A natural candidate would be $|g| = a_0\text{lub}\,|g| + a_1\text{lub}\,|g'| + \ldots$ where $a_0, a_1, \ldots$ is some sequence of positive numbers (zero cannot be allowed, for then completeness would fail). We must now concern ourselves with convergence of this infinite sum (a concern that did not arise for "d"). The idea would be to choose the a_n to go to zero sufficiently quickly with n that this sum will converge for every C^∞ g with each term finite. Unfortunately, none exists.

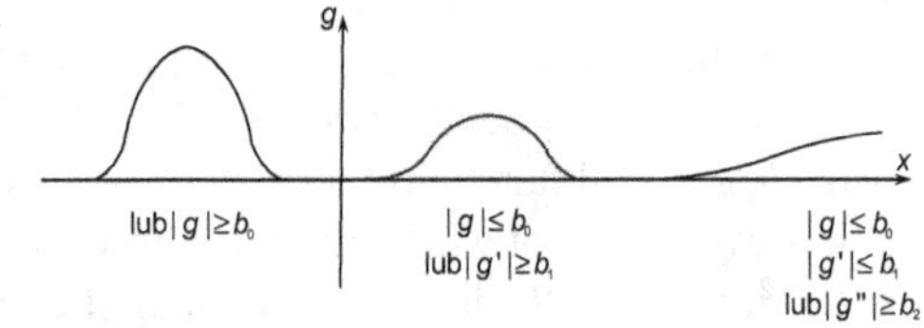

Example. Let $b_0, b_1, \ldots$ be any sequence of positive numbers. Then there exists a C^∞ function g such that each of lub $|g|$, lub $|g'|$, etc. is finite, but such that lub $|g| \geq b_0$, lub $|g'| \geq b_1$, etc. Indeed, the graph of such a g is zero except for a sequence of "bumps", where, in the first bump, the value of $|g|$ exceeds b_0; in the second, the value does not exceed b_0 but the value of the derivative exceeds b_1; in the third, the value does not exceed b_0 and the value of the derivative does not exceed b_1, but the value of the second derivative exceeds

b_2, etc.

Thus, at least the above method for making a Banach space out of C^∞ functions seems also to fail. An alternative possibility might be to try to use only certain C^∞ functions. After all, one would normally be concerned in some application only with having a "reasonable large collection" of solutions of one's equation (even though our remarks on page 119 show that there are even C^2 solutions). One might, therefore, proceed as follows. Fix positive numbers $a_0, a_1, \ldots$, and consider the collection of all C^∞ g such that $a_0 \operatorname{lub}|g| + a_1 \operatorname{lub}|g'| + \ldots$ is finite (this sum being the norm).

In this case, there is a constraint on what we can choose for the a_n: It must be true that the derivative of an admissible function (by the criterion above) is also admissible. That is to say, the a_n must be such that, if $a_0 \operatorname{lub}|g| = a_1 \operatorname{lub}|g'| + \ldots$ is finite, then $a_0 \operatorname{lub}|g'| + a_1 \operatorname{lub}|g''| + \ldots$ is also finite. But this in turn will be true if and only if the sequence of numbers whose n^{th} term is a_n/a_{n+1} is bounded. But now we are restricting ourselves essentially to analytic functions (for, given a function admissible by our criterion, expand it in Taylor series with remainder, and using boundedness a_n/a_{n+1}, show that the remainder term goes to zero as the Taylor series is made longer). It appears, therefore, that one can indeed find a suitable class of analytic functions. Unfortunately, restriction to analytic functions is too severe for, (e.g., the physical) applications of our original partial differential equations: One wants to admit at least more C^∞ functions than these.

Having now failed in several attempt to find a suitable class of functions which will form a framework for the geometrization of our partial differential equation, it may be worthwhile to just list the properties we are looking for. We seek a collection E of mappings from R to R such that:

1. Linear combinations of functions in E are in E.

2. There is a norm on E which makes it a Banach space.

3. The derivative of any function in E is in E, and "take the derivative " is a continuous (linear) mapping from E to E.

4. The set E includes a "reasonable number" of functions. Our conclusion is, then, that we have been unsuccessful in finding such a collection.

It seems reasonable, next, to see if one can find some theorem which will crystallize our failure. We give one example of a result along these lines. There exists no collection E of mappings from the open interval $(0, 1)$ to R such that i) E is a Banach space, ii) each function in E is differentiable, its derivative is in E, and "take the derivative" is a continuous mapping from E to E, iii) there is at least one function in E which is constant on $(1/2, 1)$ and which is not constant on $(0, 1)$, and iv) if g is in E and $0 \leq a \leq 1$, then the function g_a with action $g_a(x) = g(2x - a)$ is in E; furthermore, for fixed g, this mapping from $(0, 1)$ to E is C^1. The proof is quite easy. Since "take the derivative" is a continuous linear mapping from E to E, we obtain a corresponding vector field ξ on the manifold E. This vector field has unique integral curves. Let γ be such an integral curve, so, for each $t, \gamma(t) = g_t$ is an element of E. Then the statement that this is an integral curve is: $d/dtg_t = g'_t$. Let g be the function in iii) above. Then the γ with $\gamma(t)(x) = g(2x - 1 + 2t)$ is, by iv), an integral curve of our vector field. Its initial point, by iii), is a constant function. But an alternative integral curve is the constant one, $\tilde{\gamma}(t)(x) = g(2x - 1)$. By iii), these two integral curves are different. This contradiction establishes that no such collection of functions exists.

It seems, therefore, that either one must abandon the possibility of expressing partial differential equations geometrically on manifolds, or one must severely modify the rules. The first alternative would, in my view, be a severe blow to the potential applications of this subject. As to the second alternative, one could imagine at least two possibilities. On the one hand, one might look for something else (e.g., not even "functions") which one could still regard as "solutions" of our partial differential equation, and which could me made into Banach space. [The situation here is a bit reminiscent of that in Fourier analysis. The subject was a terrible mess until it was realized that one must work with the "right" set of functions – there, L^2.] One possibility which comes to mind, but which does not perhaps look too promising, is distributions. Perhaps a less severe notion of "derivative" needs to be incorporated. On the other hand, one might attempt to redo the subject of manifolds, basing them instead on topological vector spaces with less structure than Banach spaces. One possibility would be Frechet spaces (in which, essentially, the norm is replaced by a metric, but completeness is maintained). For example, the C^∞ functions with the "d" of page 122 form a Frechet space. The problem here is that it is apparently false in Frechet spaces that vector fields have unique integral curves (for the result of the previous paragraph fails if i) is replaced by "E is a Frechet space"). Perhaps there is some other kind of space in which one can work, such that reasonable classes of functions can be made into such a space, and such that, in this space, vector fields do have unique integral curves. The result of the previous paragraph, again, puts severe limitations on the possibilities.

The fact that the program is so successful and elegant for ordinary differential equations, and so appealing for partial, suggests that there should be some way to make it work.

26. Riemannian Geometry

In this section, we discuss briefly which of the various notions from Riemannian geometry in finite dimensions can be carried over to infinite.

Fix a Banach space E. A *metric* (on E – not to be confused with "metric" of "metric space") is an element g of the tensor space $\mathcal{L}(E, E; \mathbb{R})$ satisfying the following three conditions:

1. The tensor g is positive-definite. That is, for any x, y in E, $g(x, y) = g(y, x)$.

2. The tensor g is positive-definite. That is, for any nonzero x in E, $g(x, x) > 0$

3. The tensor g is invertible. First note that the tensor space $\mathcal{L}(E, E; \mathbb{R}$ is naturally isomorphic with $\mathcal{L}(E; \mathcal{L}(E; \mathbb{R})$; hence, we may (and often shall) regard g as an element of the latter. We require that there exist a g' in $\mathcal{L}(\mathcal{L}(E; \mathbb{R}); E)$ such that $g' \cdot g$ and $g \cdot g'$ are the identities on E and $\mathcal{L}(E; R)$, respectively.

Example. Let E be the Banach space of sequences of real numbers, $(r_1, , r_2, \ldots)$, the sum of the squares of whose entries converges. Let $a_1, a_2, \ldots)$ be a sequence of real numbers, and let g have the following action: For $x = (r_1, \ldots)$ and $y = (s_1, \ldots)$ in E, $g(x, y) = a_1 r_1 s_1 + a_2 r_2 s_2 + \ldots$. When is this g a metric? First note that, in order that g be well-defined, i.e., in order that the sum on the right converge for every x and y in E, it is necessary and sufficient that the $|a_i|$ be bounded. Then g is automatically symmetric. For positive-definiteness, it is necessary and sufficient that each a_i be positive. For invertibility, first identity $\mathcal{L}(E; \mathbb{R})$ with sequences of reals, so the action of $\mu = (q_1, q_2, \ldots)$ in $\mathcal{L}(E; \mathbb{R})$ on $x = (r_1, r_2, \ldots)$ in E is the real number $\mu(x) = q_1 r_1 + q_2 r_2 + \ldots$. Then, regarding g as in $\mathcal{L}(E; \mathcal{L}(E; \mathbb{R}))$, and applying to x, we have $g(x) = (a_1 r_1, a_2 r_2, \ldots)$. Clearly, then, its inverse g' must have action $g'(\mu) = (q_1/a_1, q_2/a_2, \ldots)$. But this will only be a bounded linear mapping, indeed, will only be well-defined, provided the $1/a_i$ are bounded, i.e., provided the a_i are bounded away from zero. This, then, is the condition for invertibility.

In the finite-dimensional case, the third condition above follows from the first two (i.e., in matrix terms, every finite-dimensional, symmetric, positive-definite matrix is invertible). As the example above, shows, this is not so in infinite dimensions. By the third condition, a metric establishes an isomorphism between E and its dual, $\mathcal{L}(E;\mathbb{R})$. Indeed, there is a kind of symmetry between E and its dual in this set-up. For example, g' can be regarded as an element of $\mathcal{L}(\mathcal{L}(E;\mathbb{R}, \mathcal{L}(E;\mathbb{R});\mathbb{R})$ (namely, sending (μ, ν), both in $\mathcal{L}(E;\mathbb{R})$ to $\mu(g'(\nu))$. Then this element of $\mathcal{L}(\mathcal{L}(E;\mathbb{R}), \mathcal{L}(E;\mathbb{R});\mathbb{R})$ is also symmetric, positive-definite, and invertible.

Let E be a Banach space with metric g. For x in E, set $\{x\} = (g(x,x))^{1/2}$. Then this "{ }" is actually a norm on E. [Proof: All the properties are immediate except the triangle inequality. Fix x and y in E. Then, by condition 2, we have that $g(x+ay, x+ay) = g(x,x) + ag(y,x) + ag(x,y) + a^2 g(y,y) = g(x,x) + 2ag(x,y) + a^2 g(y,y)$ must be non-negative for every number a, where we used the first condition in the second step. But this quadratic polynomial in a can be non-negative only if $(g(x,y))^2 \leq g(x,x)g(y,y)$. Now set $a = 1$ in the above, to obtain $g(x+y, x+y) \leq g(x,x) + 2(g(x,x)g(y,y))^{1/2} + g(y,y)$, where we used our inequality. Taking the square root of each side, we obtain the triangle inequality.] We claim that in fact this norm "{ }‘ is equivalent to the norm "| |" that comes with the Banach space E. Indeed, we have $\{x\} = (g(x,x))^{1/2} \leq (|g|\,|x|\,|x|)^{1/2} = |g|^{1/2}|x|$. Suppose next that the reverse inequality, $|x| \leq a\{x\}$, for some a, were false. Then we could find a sequence of vectors in $E, x_1, x_2, \ldots$ such that $|x_i| = 1$, and such that $g(x_i, x_i)$ approaches zero. Now, $|x_i| = |g'(g(x_i))| \leq |g'|\,|g(x_i)|$, whence the $|g(x_i)|$, whence the $|g(x_i)|$ must be bounded away from zero. By definition of the norm of an element (such as $g(x_i)$) of $\mathcal{L};\mathbb{R})$, therefore, there exist vectors $y_1, y_2, \ldots$ in E such that $|y_i| = 1$ and $|g(x_i)(y_i)|$ is bounded away from zero. But $|g(x_i)(y_i)|^2 = |g(x_i, y_i)|^2 \leq g(x_i, x_i)g(y_i, y_i) \leq |g|\,|y_i|^2 g(x_i, x_i) = |g| g(x_i, x_i)$, while the last expression approaches zero as i approaches infinity. This contradiction establishes that $|x| \leq a\{x\}$ for some a. We conclude that our two norms are equivalent.

A Banach space possessing an equivalent norm which arises from a metric g as above is said to be *Hilbertable.* We conclude, therefore, that a Banach space possess a metric if and only if it is Hilbertable. In this case, we might just as well use the norm "{ }", which comes from a metric, rather than the original norm "| |". A Banach space, endowed with such a norm, is called a (real) *Hilbert space.*

Which Banach spaces are Hilbertable, i.e., which possess metrics? The answer is: Very few indeed. We next amplify this remark, and also give an interesting consequence of the existence of a metric.

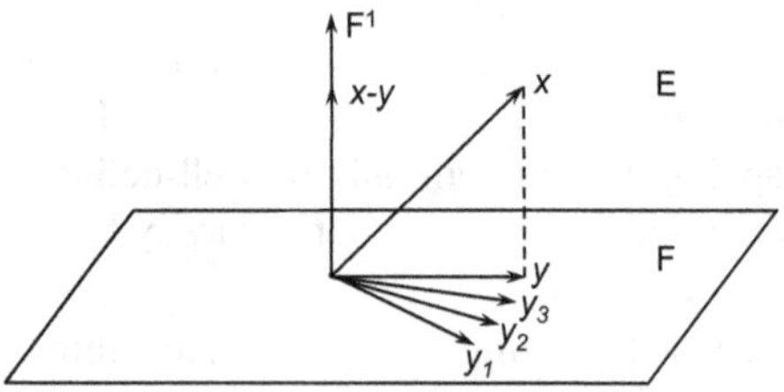

Let E be a Banach space with metric g. Two vectors, x and y, in E will be said to be *orthogonal* if $g(x, y) = 0$. Now let F be any subspace of E. Denote by $F^{\perp}$ ("F perp") the set of all vectors in E which are orthogonal to every vector in F. We claim, first, that this $F^{\perp}$ is a subspace of E. Indeed, $F^{\perp}$ is clearly a vector subspace, for any linear combination of vectors, each orthogonal to every vector in F, is itself orthogonal to every vector in F, by linearity of g. Furthermore, $F^{\perp}$ is closed. [Proof: Let $x_1, x_2, \ldots$ be in $F^{\perp}$, and let these vectors converge to x. Then, for y in F, we have, since $g(x_i, y) = 0$ and since $\lim g(x_i, y) = g(\lim \mathrm{x_i}, \mathrm{y})$, that $g(x, y) = 0$. Thus, x is also in $F^{\perp}$.] We next claim that, not only is this $F^{\perp}$ a subspace, but that it is in fact complementary to F. Fix vector x in E, and set $a = \mathrm{glb}\,|\mathrm{x} - \mathrm{y}|$, where the "glb" is over all y in F, and where we use the norm which comes from g. Let $y_1, y_2, \ldots$ be vectors in F such that $\lim |x - y_i| = a$. We have the equality $|y_i - y_j|^2 = 2|x - y_i|^2 + 2|x - y_j|^2 - 4|x - (1/2)(y_i + y_j)|^2$, which follows immediately by expanding each side using linearity of g. For i and j sufficiently large, each of the first two terms can be made as close as we wish to $2a^2$. The last term, however, is greater than or equal to $4a^2$, by definition of a. Thus, for i and j sufficiently large, the right side is as small as we wish. We conclude: The y_i form a Cauchy sequence. Let y be the vector (necessarily in F, since F is closed) to which these y_i converge. Then we have $|x - y| = a$. We next claim that $x - y$ is in $F^{\perp}$. Indeed, for any z in F and any number b, we have $a^2 \leq |x - y + bz|^2 = |x - y|^2 + 2b\, g(x - y, z) + b^2|z|^2 = a^2 + 2b\, g(x - y, z) + b^2\, |z|^2$, where the first inequality follows from the definition of a and the fact that $-y + b\,z$ is in F. Subtracting "a^2" from each side, we obtain: $0 \leq 2b\, g(x - y, z) + b^2|z|^2$. But this quadratic-polynomial inequality in b can hold for all b (as it must) if and only if $g(x - y, z) = 0$. Thus, $x - y$ is in $F^{\perp}$. But now we have $x = y + (x - y)$, an expression for x as the sum of one vector in F and another in $F^{\perp}$. That is, F and $F^{\perp}$ are complementary. [Its is almost immediate from this result that, e.g., $F^{\perp\perp} = F$.]

In particular, the above shows that, if a Banach space admits a metric, then every subspace of that Banach space split. Since we have seen examples of Banach spaces having subspaces which do not split, not every Banach space has a metric.

As usual, we now wish to pass from tensor spaces to fields. Let M be a C^p $(p \geq 1)$ manifold based on Banach space E. A *metric* (field) on M is a C^{p-1} $\mathcal{L}(E, E; \mathbb{R}$–field which is symmetric, positive-definite, and invertible. Then, for example, its inverse g' is a $\mathcal{L}(\mathcal{L}(E; \mathbb{R}); E)$–field. Thus, a metric on M gives a sort of "local distance between nearby points of M" (regarding the "infinitesimal displacement between nearby points" as being represented by a tangent vector, and the norm of this vector, defined by g, as the "distance between these two points"). A metric, then, endows M with a "local geometry". There are also various senses in which one obtains a "global geometry". We give an example.

Let M be a C^p $(p \geq 1)$ manifold based on Banach space E, and let g be a metric on M. Let γ be a curve on M, and let $a < b$ be two numbers in the interval on which γ is defined. Then, for each r in $[a, b]$, $\gamma(r)$ is a point of M, and the tangent vector to γ at r, ξ_r, is a tangent vector at $\gamma(r)$. Further, $g(\gamma(r))$ is a $\mathcal{L}(E, E; \mathbb{R})$–tensor at $\gamma(r)$. Hence, $g(\gamma(r))(\xi_r, \xi_r)$ is a non-negative number. Set $L = \int_a^b (g(\gamma(r))(\xi_r, \xi_r))^{1/2} dr$, the definite integral of one real function of a real variable. This L is called the *length* of the curve γ from a to b. [Intuitively, one "sums the infinitesimal distances between successive points along the curve from $\gamma(a)$ to $\gamma(b)$".] This definition is a reasonable one, a remark we may illustrate by the following observation. Let us reparameterize our curve, i.e., choose a smooth monotonic function f of one real variable, and set $\hat{\gamma} = \gamma \cdot f$. Let $a = f(\hat{a})$ and $b = f(\hat{b})$, so $\hat{\gamma}(\hat{a}) = \gamma(a)$ and $\hat{\gamma}(\hat{b}) = \gamma(b)$. Then, we claim, the length of γ from a to b is the same as the length of $\hat{\gamma}$ from $\hat{a}$ to $\hat{b}$. Indeed, we have $\hat{\xi}_{\hat{r}} = f' \xi_r$ (where prime denotes $d/d\hat{r}$), whence $g(\hat{\gamma}(\hat{r}))(\hat{\xi}_{\hat{r}}, \hat{\xi}_{\hat{r}}) = (f')^2 g(\gamma(r))(\xi_r, \xi_r)$. Thus, the integrals defining the length are equal by change of the independent variable. Geometrically, this is of course what one would expect.

We can now use this notion of the length of a curve to define a "global distance function" on a manifold with metric. Let M, p, E, and g be as usual, and suppose further that M is connected, i.e., has the property that any two of its points may be joined by some curve. Given any two points, p and p', of M, set $d(p, p)$ the greatest lower bound of the length of curves joining p and p'. This "d", we claim, is a metric. [Proof: That $d(p, p') \geq 0$, and that $d(p, p') = d(p', p)$, are immediate. That $d(p, p') + d(p', p'') \geq d(p, p'')$ is also clear, since one curve from p to p'' is obtained by first tracing a curve from p to p', and then from p' to p''. We have only to show, therefore, that $d(p, p') = 0$ implies $p = p'$. Introducing a chart containing p, it suffices to do this within the Banach space E. Since, furthermore, the metric field is continuous, it suffices to consider the case when the metric is constant (for one can always find a "lower bound metric" in some neighborhood of p). Thus, we must show: For E a Banach space, g a metric on E, and p and p' distinct points of E, $d(p, p') > 0$. Let f be the function on E which assigns to x in E $f(x) = (g(x - p, x - p))^{1/2}$. Then, along any curve from p, f increases more slowly than the length of the curve increases (since the rate of change of f is measured by the norm of the component of the tangent vector in the direction away from p while the rate of change of length is measured by the norm of the tangent vector). Hence, $d(p, p') \geq (g(p' - p, p' - p))^{1/2}$.] Thus, a manifold-with-metric has the structure of a metric space.

The discussion above is a brief summary of "the geometry of manifold-with-metrics". We conclude this section with a discussion of metrics from a more algebraic point of view.

One of the most important and useful properties of a metric (field) is that it leads to a unique derivative operator. More precisely, we have: Let

M be a C^p ($p \geq 2$) manifold based on Banach space E, with C^{p-1} metric g. Then there is one and only one derivative operator ∇ on M satisfying $\nabla g = 0$. The proof is computational and completely straightforward: One writes down the answer. Let U, ψ be a chart, and $\hat{g}$ the component of g. For u, v, and w in E, consider $1/2(D\hat{g}(u)(v, w) + D\hat{g}(v)(u, w) - D\hat{g}(w)(u, v))$. Fixing u and v, this assigns, to each w in E, a real number, i.e., is an element of $\mathcal{L}(E; \mathbb{R})$. Applying $\hat{g}'$ to this mapping, therefore, we obtain an element of E, an element which depends, of course, on our choices of u and v. Write this element $\Gamma(u, v)$, so Γ is an element of $\mathcal{L}(E, E; E)$. We now claim: This Γ is a connection. [Sketch of proof: Change the chart. We know the change in the component of g, and hence can compute, from the formula above, the change in Γ. The result is precisely the formula for the change in the component of a connection under chart-change.] Thus, we obtain a derivative operator on M. This ∇ further satisfies $\nabla g = 0$, for this equation, written in terms of a chart, is precisely the formula above used to define ∇.

Thus, once one has a manifold with metric, one automatically has a derivative operator, and hence write differential equations on tensor fields on M.

We remark that, just as in the finite-dimensional case, one can define the curvature tensor and obtain its symmetries (although one has neither the Ricci tensor, the conformal tensor, nor the scalar curvature, since these require a "contraction" which is not available in infinite dimensions), geodesics, etc.

Appendix

Problems

1. Find an example of Banach space E, vector subspace A of vector space E, such that A is finite co-dimension but not closed.

2. Let E be a Banach space. Show that every finite-dimensional subspace of vector space E is closed.

3. Find an example of a vector space, and two norms thereon each of which makes this vector space a Banach space, such that these norms are not equivalent.

4. Show that a subspace of a Banach space containing a ball is the entire Banach space.

5. Let E be a Banach space. Find all subspaces of E having a unique complementary subspace.

6. By a basis for Banach space E, we mean a sequence $x_1, x_2, \ldots$ of vectors in E such that, given any vector x in E, there is one and only one sequence $a_1, a_2, \ldots$ of real numbers such that the sequence in E whose n^{th} term is $y_n = a_1 x_1 + \ldots + a_n x_n$ converges to x. [That is, every element of E can be written as an "infinite linear combination" of the x_i.] Show that our examples of Banach spaces have bases. Find an example of Banach space without basis.

7. Find a vector space having no norm which makes it a Banach space.

8. Let E and F be Banach spaces, and let T be a linear mapping (of vector spaces) from E to F. Denote by A the subset of $E \times F$ consisting of elements of the form $(x, T(x))$, with x in E. Prove that A is a vector subspace. Show that T is bounded if and only if A is closed in $E \times F$. [Hint: Open mapping theorem.]

9. Let E consist of all sequences, $(r_1, r_2, \ldots)$ of real numbers with $|r_2 - r_1| + |r_3 - r_2| + \ldots$ finite. Let the norm of such a sequence be this sum, plus $|r_1|$. Prove that this E is a Banach space.

10. Show that a Banach space is finite-dimensional if and only if every subspace is closed.

11. Define the product of an infinite number of Banach spaces.

12. Give an example to show that the open mapping theorem is false for normed vector spaces (i.e., without completeness).

13. Prove that every subspace of the Banach space E^2 (page 19) splits.

14. Prove that, if every subspace of E splits and every subspace of F splits, then every subspace of $E \times F$ splits.

About the author

Robert Geroch is a theoretical physicist and professor at the University of Chicago. He obtained his Ph.D. degree from Princeton University in 1967 under the supervision of John Archibald Wheeler. His main research interests lie in mathematical physics and general relativity.

Geroch's approach to teaching theoretical physics masterfully intertwines the explanations of physical phenomena and the mathematical structures used for their description in such a way that both reinforce each other to facilitate the understanding of even the most abstract and subtle issues. He has been also investing great effort in teaching physics and mathematical physics to non-science students.

Robert Geroch with his dog Rusty

www.ingramcontent.com/pod-product-compliance
Lightning Source LLC
Chambersburg PA
CBHW062024200326
41519CB00017B/4911
9781927763155